Palgrave Studies in World Environmental History

Series Editors

Vinita Damodaran
Department of History
University of Sussex
Brighton, United Kingdom

Rohan D'Souza
Graduate School of Asian and African Area Studies
Kyoto University
Kyoto, Japan

Sujit Sivasundaram
University of Cambridge
Cambridge, United Kingdom

James John Beattie
Department of History
University of Waikato
Hamilton, New Zealand

The widespread perception of a global environmental crisis has stimulated the burgeoning interest in environmental studies. This has encouraged a wide range of scholars, including historians, to place the environment at the heart of their analytical and conceptual explorations. As a result, the understanding of the history of human interactions with all parts of the cultivated and non-cultivated surface of the earth and with living organisms and other physical phenomena is increasingly seen as an essential aspect both of historical scholarship and in adjacent fields, such as the history of science, anthropology, geography, and sociology. Environmental history can be of considerable assistance in efforts to comprehend the traumatic environmental difficulties facing us today, while making us reconsider the bounds of possibility open to humans over time and space in their interaction with different environments. This new series explores these interactions in studies that together touch on all parts of the globe and all manner of environments including the built environment. Books in the series will come from a wide range of fields of scholarship, from the sciences, social sciences, and humanities. The series particularly encourages interdisciplinary projects that emphasize historical engagement with science and other fields of study.

More information about this series at
http://www.springer.com/series/14570

Lill-Ann Körber • Scott MacKenzie • Anna Westerståhl Stenport
Editors

Arctic Environmental Modernities

From the Age of Polar Exploration to the Era of the Anthropocene

Editors
Lill-Ann Körber
Department of Linguistics and Scandinavian Studies
University of Oslo, Oslo, Norway

Scott MacKenzie
Department of Film and Media
Queen's University, Kingston, Ontario, Canada

Anna Westerståhl Stenport
School of Modern Languages
Georgia Institute of Technology, Atlanta, Georgia, USA

Palgrave Studies in World Environmental History
ISBN 978-3-319-81821-4   ISBN 978-3-319-39116-8   (eBook)
DOI 10.1007/978-3-319-39116-8

Cover image © ARCTIC IMAGES / Alamy Stock Photo

Printed on acid-free paper

This Palgrave Macmillan imprint is published by Springer Nature
The registered company is Springer International Publishing AG
The registered company address is: Gewerbestrasse 11, 6330 Cham, Switzerland

Acknowledgments

The editors are grateful to the many people, organizations, and granting agencies that have supported our work on this project. We benefited from the opportunity to organize multi-year Arctic programs at the Society for the Advancement of Scandinavian Study's annual conferences between 2013 and 2016, a project that originated in San Francisco in 2013. The early stages of research that went into this book were presented at the 2014 "Arctic Modernities" conference in Tromsø, Norway, and in conversation with Anka Ryall, Henning Howlid Wærp, and Johan Schimanski of the Arctic Modernities and Arctic Discourses projects at UiT The Arctic University of Norway. We value the feedback we received at the Hammarskjöld lecture series at the Nordeuropa-Institut at Humboldt-Universität in Berlin in 2015. Lill-Ann Körber wishes to acknowledge the support provided by the Henrik Steffens Professorship at Humboldt-Universität and the Deutscher Akademischer Austauschdienst; Scott MacKenzie, the Fund for Scholarly Research and Creative Work and Professional Development, Queen's University; Anna Westerståhl Stenport, the European Union Centre of Excellence, the Jean Monnet Center of Excellence, the Conrad Professorial Humanities Fund, and the Research Board at the University of Illinois at Urbana-Champaign; and Scott MacKenzie and Anna Westerståhl Stenport wish to acknowledge the Social Sciences and Humanities Research Council of Canada, which provided them with an Insight Grant. We are grateful for the diligent work undertaken by our research assistants Noelle Belanger,

Carlo Di-Gioulio, Paul Greiner, and Garrett Traylor, and for the editorial assistance provided by Angela Anderson. We benefited from the astute feedback offered by colleagues Verena Höfig, Janke Klok, Mark Safstrom, and Stefanie von Schnurbein, as well as by our external readers.

Contents

Notes on Contributors

Dag Avango is a researcher at the Division of History and Technology at KTH Royal Institute of Technology, Sweden. His scholarship addresses the relationship between scientific research, natural resource exploitation, and geopolitics in the polar regions (Arctic and Antarctic); his complementary interests include industrial heritage studies. Situated at the interface between archaeology, archival methods, and history inquiry, Avango's work spans both academic and public sector interests. He is the lead instructor for the interdisciplinary and international summer course "Environment and Society in a Changing Arctic".

Lisa E. Bloom is a visiting scholar at the Center for the Study of Women at UCLA. Bloom has written extensively on the polar regions, feminist cultural studies, visual culture, film, and contemporary art. She is the author of *Gender on Ice: American Ideologies of Polar Expeditions* (1993), the first critical feminist cultural studies book on the polar regions. More recently, she has written numerous articles on the polar regions, contemporary art, and film. She is currently writing a book titled *Contemporary Art and Climate Change in the Polar Regions: Gender After Ice* (2017). The book examines aspects of feminist and environmentalist art and film in relation to new scholarship of the polar regions, bringing together issues routinely kept apart in climate change debates such as connecting gender to nationalism, capitalism, and post-colonialism.

Ann-Sofie Nielsen Gremaud holds a PhD in visual culture and is a postdoctoral fellow at the University of Copenhagen and guest researcher at the EDDA Center of Excellence at Háskóli Íslands in Reykjavík. She is part of the international joint research project "Denmark and the New North Atlantic" and has published on cryptocolonialism, environmental discourses in art and politics, the making of Arctic Iceland, and geographies of crisis in Icelandic photography.

Kikki Jernsletten is a Sámi of Tromsø, Norway. She was born in 1968, has four children and is a grandmother of four. She is a singer/songwriter (aka Kikki Aikio), a professionally trained masseuse, a journalist and author, and holds a PhD.

Lilya Kaganovsky is Associate Professor of Slavic and Director of the Program in Comparative and World Literature, at the University of Illinois at Urbana-Champaign. Her books include: *How the Soviet Man Was Unmade* (2008); *Mad Men, Mad World: Sex, Politics, Style and the 1960s* (with Lauren M. E. Goodlad and Robert A. Rushing, 2013); and *Sound, Music, Speech in Soviet and Post-Soviet Cinema* (with Masha Salazkina, 2014).

Lill-Ann Körber holds a PhD in Scandinavian Studies from Humboldt-Universität zu Berlin, Germany. She is a post-doctoral fellow at the University of Oslo in the research project "Scandinavian Narratives of Guilt and Privilege in an Age of Globalization" and Associate Professor II of Modern Scandinavian Literature at the University of Bergen, Norway. Her research interests include Greenlandic contemporary culture, Greenland and the Arctic in literature and film, and Africa and the "Black Atlantic" in Scandinavian literature, art, and memory culture. She co-directed the Greenland Eyes International Film Festival in Berlin in 2012. Recent works include her doctoral thesis *Badende Männer. Der nackte männliche Körper in der skandinavischen Malerei und Fotografie am Anfang des 20. Jahrhunderts* (2013), the article "'See the crashing masses of white death ...' Greenland, Germany and the Sublime in the Bergfilm *SOS Eisberg*" (in Scott MacKenzie and Anna Westerståhl Stenport (eds.) *Films on Ice: Cinemas of the Arctic*, 2015), and *The Postcolonial North Atlantic: Iceland, Greenland and the Faroe Islands* (co-edited with Ebbe Volquardsen, 2014).

Scott MacKenzie teaches in the Department of Film and Media at Queen's University, Canada. His articles have appeared in numerous journals, including *Screen, Camera Obscura, Moving Image*, and *Public*. His books include: *Cinema and Nation* (with Mette Hjort, 2000); *Purity and Provocation: Dogma '95* (with Mette Hjort, 2003); *Screening Québec:*

Québécois Moving Images, National Identity and the Public Sphere (2004); *The Perils of Pedagogy: The Works of John Greyson* (with Brenda Longfellow and Thomas Waugh, 2013); *Film Manifestos and Global Cinema Cultures* (2014); *Films on Ice: Cinemas of the Arctic* (with Anna Westerståhl Stenport, 2015); *Arctic Cinemas and the Documentary Ethos* (with Lilya Kaganosky and Anna Westerståhl Stenport, 2017); *The Cinema, too, Must be Destroyed: The Films of Guy Debord* (forthcoming), and *Process Cinema: Handmade Film in the Digital Age* (with Janine Marchessault, forthcoming).

Kristian H. Nielsen is Associate Professor at the Centre for Science Studies at Aarhus University, Denmark. His main research interests are the history of science and technology and science communication. He has published in journals such as *Annals of Science, British Journal for the History of Science, Centaurus, Environmental Communication, Historical Studies of the Natural Sciences, New Global Studies, Public Understanding of Science, Science as Culture*, and *Science Communication*.

Torill Nyseth holds MA and PhD degrees in planning. She is Professor at the Department of Sociology, Political Science and Community planning at UiT The Arctic University of Norway, Tromsø. Her research fields are in local democracy, network governance, place development, and urban planning. She has published many journal articles in political science, entrepreneurship, urban planning, and governance in *Local Government Studies, Planning Theory, Planning Theory and Practice, Cities, Acta Borealia*, and *European Urban and Regional Planning*. She has also published *Place Reinvention: Northern Perspectives* (co-edited with Arvid Viken, 2009).

Peder Roberts is a researcher in the Division of History of Science, Technology and Environment at KTH Royal Institute of Technology. He was educated at the University of New South Wales and Stanford University and has previously worked at the University of Strasbourg. He is the author of *The European Antarctic: Science and Strategy in Scandinavia and the British Empire* (Palgrave, 2011) and co-edited with Simone Turchetti *The Surveillance Imperative: Geosciences During the Cold War and Beyond* (Palgrave, 2014). His research interests cover the history and politics of the polar regions (particular in terms of science and the environment) and the relationship between science and geopolitics during the Cold War.

Mark Safstrom is Lecturer of Swedish and Scandinavian Studies at the University of Illinois at Urbana-Champaign. His primary research interests are in nineteenth and early twentieth-century Scandinavian history, with emphasis on the so-called "folk movements" (the temperance movement, labor movement, and the religious revivals), as well as the history of Scandinavian immigration to North America, and Scandinavian polar explorers and their travel accounts. Safstrom regularly teaches on the KTH-Illinois "Stockholm Summer Arctic Program: Environment and Society in a Changing Arctic," which has field site components in Svalbard, Norway and Northern Sweden.

Sverker Sörlin is Professor in the Division of History of Science, Technology and Environment at the KTH Royal Institute of Technology, Stockholm, and is co-founder of the KTH Environmental Humanities Laboratory. His research is focused on the role of knowledge in modern societies and the formation of environmental expertise. He is the editor of *Narrating the Arctic: A Cultural History of Nordic Scientific Practices* (with Michael Bravo, 2002), *Science, Geopolitics and Culture in the Polar Region: Norden beyond Borders* (2013), *Northscapes: History, Technology, and the Making of Northern Environments* (with Dolly Jørgensen, 2013), and *The Future of Nature: Documents of Global Change* (with Libby Robin and Paul Warde, 2013).

Anna Westerståhl Stenport is Professor and Chair of the School of Modern Languages, Georgia Tech. She is co-editor of *Films on Ice: Cinemas of the Arctic* (with Scott MacKenzie, 2015), the first book on transnational global Arctic film and media, and *Arctic Cinemas and the Documentary Ethos* (with Lilya Kaganovsky and Scott MacKenzie, forthcoming). She has written extensively on contemporary film and media (including digitality studies, adaptation, transnationalism, and indigenous cinemas) and her work has appeared in *Comparative Literature, Cinema Journal, Convergence, Camera Obscura, Film History, Modernism/ Modernity, Public, Journal of Scandinavian Cinema* and many other journals and edited books. Her books in English include *Nordic Film Classics: Lukas Moodysson's "Show Me Love"* (2012), *Locating August Strindberg's Prose: Modernism, Transnationalism, and Setting* (2010), and *The International Strindberg: New Critical Essays* (ed., 2012).

Troy Storfjell is Associate Professor of Norwegian and Scandinavian Studies at Pacific Lutheran University, where he also chairs the Native

American and Indigenous Studies Working Group. He earned his PhD at the University of Wisconsin (2001) and has also held a postdoctoral appointment at the University of Tromsø. He has published several articles on Knut Hamsun, and his current work is in the field of indigenous methodologies.

Andrew Stuhl is a Assistant Professor of Environmental Humanities at Bucknell University. His teaching and research sits at the crossroads of environmental history, the history of science, and environmental studies. His most recent project, *Unfreezing the Arctic: A Century of Science, Intervention, and Environmental Transformation* (2016), traces the recurring attempts to study, exploit, and protect the North American Arctic since the 1850s to illuminate the historical legacies at stake in modern climate change and globalization. He is also interested in the histories of pipelines, laboratories, and environmental impact assessments in the Arctic.

Eva-Maria Svensson is Professor of Law at the Department of Law, University of Gothenburg, Sweden. Her main interests are legal philosophy and theory, particularly in the field of feminist/gender legal studies. One of her contemporary research collaborations is the international cross-disciplinary comparative research program, TUARQ (Tromsö, Umeå, Archangelsk, Rovaniemi, and Queen's and Québec [Université Laval] Universities) Network on Gender Equality in the Arctic, with scholars from law, economical statistics and political science who have competence in gender studies. The focus of the program is on public mechanisms for gender equality in the Arctic.

Kirsten Thisted is Associate Professor at the Minority Studies Section, Department of Cross-Cultural and Regional Studies, University of Copenhagen. She has published on colonial encounters in Danish literature, and on Greenlandic oral tradition and modern literature and culture. She is currently heading the research project "Denmark and the New North Atlantic." This project concerns the renegotiations of identities taking place in all parts of the North Atlantic these days, as a result of independence processes, climate change, and globalization. The project is funded by the Carlsberg Foundation.

Synnøve Marie Vik is a PhD candidate at the Department of Information Science and Media Studies at the University of Bergen, Norway. Her thesis explores the relationship between nature, technology and their media ecologies in contemporary visual culture. She is a critic and a curator, currently working as an Art Adviser to the City Council of Bergen. She has also been a Fulbright Scholar in the Department of Art History and Archeology at Columbia University.

Nina Wormbs   holds an MSc in engineering physics and a PhD in the history of technology. She is Associate Professor of the History of Science and Technology and serves as Head of the Division of History of Science, Technology and Environment at KTH Royal Institute of Technology. Apart from an interest in how Arctic futures are construed at present and historically, she has written on media and technological change during the twentieth century with a special focus on infrastructure and politics. She has had several commissions from the Swedish government and serves on the board of the Nobel Museum. Recent publications include *Media and the Politics of Arctic Climate Change: When the Ice Breaks* (co-edited with Miyase Christensen and Annika E. Nilsson, Palgrave Macmillan, 2013).

LIST OF FIGURES

Introduction: Arctic Modernities, Environmental Politics, and the Era of the Anthropocene

Lill-Ann Körber, Scott MacKenzie,
and Anna Westerståhl Stenport

Since its "discovery," the Arctic has held a longstanding significance as a critical and exceptional space of modernity. It has been utilized and imagined as a location where the past, present, and future of the planet's environmental and geopolitical systems are played out. These imaginations and projections have hit a crescendo in recent years, catalyzed by anthropogenic climate change, accelerating resource extraction, mass tourism, and a heightened global awareness and activism regarding environmental

L.-A. Körber (✉)
Department of Linguistics and Scandinavian Studies, University of Oslo, Oslo, Norway

S. MacKenzie
Department of Film and Media, Queen's University, Kingston, ON, Canada

A.W. Stenport
School of Modern Languages, Georgia Institute of Technology, Atlanta, GA, USA

© The Author(s) 2017
L.-A. Körber et al. (eds.), *Arctic Environmental Modernities*,
DOI 10.1007/978-3-319-39116-8_1

1

change, Indigenous rights, and nature preservation. *Arctic Environmental Modernities* critically investigates the exceptional status of Arctic environmental discourses and practices by foregrounding the diversity, hybridity, and multiplicity of Arctic modernities, and by nuancing differentiations between sublime "nature," cultural and vernacular landscapes and cityscapes, and social practices. To this end, the book addresses the rise and conflicted status of Arctic modernities from nineteenth-century European exploration of the Arctic to the present day. *Arctic Environmental Modernities* provides a framework for examining the continuing role of the explorer mythology in accounts of Arctic modernities, while foregrounding methodologies that contest such a monolithic historiography.

ARCTIC EXPLORATION AND MODERNITY

Arctic exploration and its relationship to questions of modernity, in this context, must be seen through a specific frame: that of the European and North American explorers who went there and then either left or died there, with the goal of "expanding" territorial holdings in the name of the nation state. The Arctic explorer myth assumes that there is no "staying," yet the generations of offspring that emerge from the era of exploration tell another story, as does the environmental impact of continuous exploration in the global circumpolar North. The book examines the history of Arctic exploration to question the construction of this myth and the counternarratives that have been used to challenge its centrality to discourses of European modernity. European modernity, *pace* Marshall Berman and Eric Hobsbawm, begins in 1789 (Hobsbawm 1962; Berman 1988). This period of political and industrial revolutions meant the start of an era in need of ever greater supplies of fossil fuels (including oil from whaling in the Arctic), and the results are seen in today's baseline of disproportionate environmental effects in the far North. There is a direct genealogy from industrialization and its social formations to the two interrelated challenges facing the Arctic presently: climate change (warming environments) and the continuous drive for resource extraction (with multifarious environmental impacts). Most European assumptions of Arctic modernity arguably begin nearly a century after the emergence of the Enlightenment in continental Europe, as European and North American exploration of the polar regions and the colonization of the North accelerated in the mid-to-late 1800s. This era of polar exploration can be understood as an attempt to mobilize Western technology and the global expansion of territorial holdings, by which the colonial practices of nation states (from cartography to whaling) were further sanctioned. More contemporary accounts of Arctic modernities place nature and the environment back at the center of discourses of modernity, especially in relation to fossil fuel and rare earth mineral resource extraction, the politics governing marine life and fishing, climate change, pollution, Indigenous rights, and the ideological belief systems that underpin questions of geopolitical sovereignty (Bravo and Triscott 2011; Dodds and Nuttall 2015; Kjeldaas and Ryall 2015).

Arctic Environmental Modernities investigates how a study of the Arctic region as a privileged site of modernity articulates globally significant but often overlooked intersections between environmentalism and sustainability, Indigenous epistemologies and representational practices, decolonization strategies, and governmentality (especially in the Nordic and Canadian welfare states). The book offers a pan-Arctic scope. Most of the constituent nation states are addressed in historical and contemporary frames. Attention is also paid to the distinct political, judicial, cultural, colonial, and sociological aspects of these Arctic states. For instance, the book avoids totalizing Scandinavia as a homogeneous region, and foregrounds discrete political and cultural formations with contrasting histories; this is especially important, as one of the main arguments put forth in this book is that the Arctic itself is in no way homogeneous. Contrasting discourses are evident in a range of representational practices in the arts, cinema, ethnography, and literature, as well as in corporate, government, NGO, and scientific documentation. These varied and conflicting discourses are critical to understanding how theories of modernity are articulated, implemented, and occasionally rejected in the Arctic regions and beyond.

Arctic Environmental Modernities additionally examines narratives of Arctic counter-modernities that complicate political, social, and discursive assumptions of European and Western models. For example, as many of the chapters in this book demonstrate, the Arctic region has functioned as a space to project Euro-American modernity and modernization ideologies, while simultaneously challenging these paradigms from within, constructing location-specific concepts of modernity. Furthermore, "the Arctic" was conceived as a space where ideas of a nostalgic past, or utopian or dystopian futures, could be put to the test. The book therefore builds on notions of "alternative modernities" (Gaonkar 2001) and "multiple modernities" (Eisenstadt 2000) as well as on approaches that recognize Indigenous or vernacular cosmopolitanisms (Bhabha 1996; Werner

2006; Forte 2010). Shmuel Eisenstadt affirms that "modernity and Westernization are not identical," but that we in the twenty-first century are facing a "multiplicity of continually evolving modernities" (Eisenstadt 2000: 2f). Dilip Parameshwar Gaonkar also argues that "modernity today is global and multiple and no longer has a governing center of master narratives to accompany it," especially at "a time when non-Western people everywhere begin to engage critically their own hybrid modernities" (Gaonkar 2001: 14). Following these lines of thought, *Arctic Environmental Modernities* is therefore dedicated to an exploration of the ways in which such approaches pertain to pluralist conceptualizations of the Arctic. Indeed, the need for counter-modernities is cogently argued for by Edward Said: "what does need to be remembered is that narratives of emancipation and enlightenment in their strongest form were also narratives of integration and not separation, the stories of people who had been excluded from the main group but who were now fighting for a place in it" (Said 1994: xxvi). This book also analyzes gendered constructions of "the Arctic," complicating the region as a bastion for the discursive construction of heteronormative masculinities as ruling over nature (see Bloom 1993; Hill 2008; MacKenzie and Stenport 2013). The Arctic has continuously been an arena for the performance of conflicted narratives about masculine heroism, supposedly anchored through recourse to normative male rationality and beliefs in technological progress. Foregrounding alternative Arctic modernities, which question colonial, gendered, capitalist, and racialized power structures, is thus central to the book's argument.

Arctic Environments

Arctic Environmental Modernities thereby draws upon a diverse array of definitions of "the Arctic" beginning in the late 1800s and the era of polar exploration, which are often in conflict with one another (see Ryall et al. 2010). For instance, the concept of the Anthropocene, increasingly promulgated through the media and factions of the scientific community during the last decade, has substantial bearing on the politics of representing the Arctic in the twenty-first century. First developed by ecologist Eugene F. Stoermer in the 1980s, and refined by atmospheric chemist Paul Crutzen, the Anthropocene addresses the vastly accelerated rate of climate change brought on by humans: "our species' whole recorded history has taken place in the geological period called the Holocene—the brief interval stretching back 10,000 years. But our collective actions have brought us into uncharted territories. A growing number of scientists think we have entered a new geological epoch that needs a new name— the Anthropocene" (www.theanthropocene.info; see also Robin 2013). While the notion of the Anthropocene is contested (see, for instance, Malm and Hornborg 2014 and Chernilo 2016), both in terms of when it began and whether it exists, we mobilize the term as a means by which to frame the environmental modernity that is the *de facto* Arctic. That includes the vast ecological, ideological, and political changes that have emerged since the Industrial Revolution. The Anthropocene allows for tracing the confluences between industrialization, resource extraction, advanced capitalism, and neoliberalism to understand these developments as not discrete and unrelated phenomena. Therefore, many of the book's chapters implicitly and explicitly engage with the Anthropocene in the Arctic context, addressing local and global implications of environmental change and environmentalist thought.

Arctic Environmental Modernities demonstrates how various definitions of the Arctic are always ideological, mobilized by various actors to their own ends. The consequences of ongoing enviro-spatial shifts in the Arctic can seem paradoxical: we encounter simultaneous processes of regionalization, localization, Indigenization, globalization, and nationalization. The plethora of competing definitions is a constitutive part of Arctic environmental modernities, reflecting different perceptions of what environments are in relation to the human and social cultures and ideologies that shape them, and by which environments simultaneously inform assumptions of modernity—in the Arctic and elsewhere. To this end, many different notions of the Arctic are mobilized in the book, foregrounding the contested nature of its constitution, its environmental implications, and how it is always discursively constructed.

Most contemporary definitions of the Arctic are "environmental," originating in supposedly empirical, observable, and quantifiable parameters of the natural world: climate (the 10 °C July isotherm); vegetation (the tree line); marine boundary (temperature and saline quotient of ocean water); and cartographic (the Arctic Circle at 66° 32' north) in ways that roughly account for the northernmost areas of the eight Arctic Council nation states: USA (Alaska), Canada, Denmark (Greenland), Iceland, Norway, Sweden, Finland, and Russia. These positivist parameters, which map onto the geopolitical priorities of sovereign states, are not as empirically stable as one might first assume. As the Arctic is a negotiated region consisting of environments, cultures, histories, practices, and modernities—permeated by geopolitical tensions—the definition offered by the Arctic Council working group Arctic Monitoring and Assessment Programme (AMAP) offers a contrasting account, tied to questions of sovereignty and ideology that are also shaped by specific environments. AMAP's definition indeed foregrounds the region's constructed constitution: "AMAP has established a circumpolar region as a focus for its assessment activities that includes both High Arctic and sub-Arctic regions." This "established" assessment area reflects key components of AMAP's charge to monitor pollutants, assess evidence and impact of climate change, and promote socio-economic development. These aspects are key to understanding the environmental modernities that have shaped and continue to impact the region, and the understanding of it in other parts of the world. AMAP's assessment protocol thus effectively constitutes "the Arctic" as based on "collaborations with relevant groups," with a primary purpose being to "answer the needs of policy-makers" (AMAP 1997). This conceptualization of "the Arctic" in effect promotes regional environmental, policy, and socio-economic factors as constitutive—which is clear from the Arctic Council's activities—while foregrounding their discursive and negotiated status. AMAP, like all definitions, is not without its problems; see Wormbs and Sörlin, in this volume, for a critical appraisal of the limitations of the AMAP definition.

The Inuit Circumpolar Council (ICC), in contrast, defines the Arctic by way of the international solidarity that exists between Inuit Indigenous peoples: "to thrive in their circumpolar homeland, Inuit had the vision to realize they must speak with a united voice on issues of common concern and combine their energies and talents towards protecting and promoting their way of life" (ICC). The ICC does not address the transnational and bilateral relationships between states; instead, it envisions a shared and interconnected space that needs to be defended through, among other things: "strengthen[ing] unity among Inuit of the circumpolar region; [and] promot[ing] Inuit rights and interests on an international level" (ICC).

The Arctic has repeatedly been defined by historical and nationalist accounts from outside the region, and not by Indigenous and local knowledge. For instance, the Sámi tend not to use the term "Arctic" to describe Sápmi, the transnational region of Northern Fenno-Scandinavia and the Murmansk Peninsula; in Sámi, it is an outsider's term. While many inhabitants do not take the term as a definition of where they live, this lack of identification does not mean that inhabitants are not affected by the policies and practices mobilized under the term "the Arctic." Some scholars try to solve the dilemma that surrounds the term by replacing it with "the North." Tim Ingold, for instance, has argued that this term ought to be used in place of "the Arctic," as the former has both conceptual and geographical value, while the latter is a homogenizing term imported from the outside (Ingold 2013: 37–48; see also Keskitalo 2004, 2009). But the use of "the North" also risks essentializing a region, locating it as somehow at the far spectrum of the compass, as if dislodged from both material and discursive frameworks. At the same time, there are pertinent reasons why both histories of "the Arctic" and "the North" are so problematic. As Dolly Jørgensen and Sverker Sörlin argue: "a general history of the North, or of the circumpolar regions, is yet to be found. The reason is not hard to see: apart from the old stereotype of emptiness and silence, the global North has not until quite recently, been a region in its own right" (2013: 4; for historical treatments of the Arctic as a region, see Emmerson 2010; McCannon 2012; McGhee 2007). Therefore, our own use of the term "the Arctic" is strategic, as it speaks to the way in which outsiders through policy, politics, and aesthetics have defined the region. This dialectic between "inside" and "outside" Arctic environmental modernities foregrounds the contested nature of the region's imaginary and its role in geopolitics.

This book brings together an international and interdisciplinary group of scholars, representing fields as diverse as legal and policy studies, environmental humanities, gender studies, critical ethnography, art history, film and media studies, comparative literature, Indigenous studies, religious studies, and the history of science and technology. The book's contributors address a series of questions, including: How does the Arctic, as one of the last potential spaces on the planet ripe for colonialization and exploitation, become reimagined in the processes that multinational corporations and nation states deploy to transform it into an infinitely exploitable resource? How do these accounts mobilize the contentious notions of Arctic exploration and Arctic modernity of the nineteenth century? Do new technologies engender new Arctic imaginaries of modernity, and how are these imaginaries mobilized to new or recurrent political, ideological, or environmental ends? How do these discourses rely on and challenge a long history of environmental and modernist imaginaries of the global North? Are there particularities that help elucidate the power and function of how the Arctic region continues to engage the dialectic between fear and fascination that spans environmental considerations whether past, present, or future? Which actors and institutions, from Indigenous populations and self-governing bodies to traditional nation states, are involved in shaping and imagining the ecologies of the Arctic today? Which continuities and discontinuities can be observed regarding traditions of representing and imagining Arctic modernities? Who owns the sovereignty of information and interpretation of these ongoing processes? And lastly, what kind of tensions and contradictions are at play in these forms of cultural production?

WHOSE ARCTIC?

Even in twenty-first-century environmental humanities, Indigenous and local knowledge, with all its cultural, historical, and technological richness, is often marginalized: the strategic and ideological goals of Western modernity and "progress" remain superimposed onto discourses and practices of the Arctic. And while the popularity of Arctic studies is increasing, and an increasing number of Indigenous and local scholars shape the field, it was outsider perspectives that dominated research during the twentieth century (for a salient critique of this form of intellectual colonialism, see Tuhiwai Smith 1999). A survey of scholarly articles, published in *Arctic*, a leading multidisciplinary journal, indeed showed that scholarship about the Arctic reflected mostly "'southern interests,' (Harrison and Hodgson 1987: 330) for one example, non-renewable resources, militarism and sovereignty" (Keskitalo 2009: 30). This is not to argue in any way that one form of "Indigenous knowledge" can simply stand in for Western modernity; the question of locality, as noted above, is central. For instance, Indigenous scholarship looks very different in Sápmi and in Nunavut, and, of course, varies within these locations as well. Rauna Kuokkanen shows the significance of Sámi revitalization and political mobilization as evolving in very close connection with assumptions of the Nordic welfare state, including its forms of parliamentary representation, emphasis on consensus culture, and long-term strategies of subsuming political struggle for gender equality as a subset of welfare state ideology and politics (Kuokkonen 2011). In Sápmi, there is a long history of emphasizing the significance of language policies and cultural revitalization, but as of yet no coherent strategy for independence or self-governance (e.g., Kuokkanen 2009). In Sápmi, conflicts regularly arise about ownership of the land, raising questions as to when national or international law applies. The ILO-169 convention, regulating the protection of the rights of Indigenous peoples, has, of the four nation states with a Sámi population, so far only been ratified by Norway (of the other Arctic states with Indigenous populations, it has been ratified by Denmark, but not by Canada, Russia, or the USA). In Sweden and Finland the Sámi are classified as a minority language group, not as Indigenous populations. Russia's stance has been, for a long time, to disregard any status of Sámi as an Indigenous population, just as the suppression of Northern ethnic minorities has been consistently harsh for centuries (Slezkine 1994). Recently, Vladimir Putin sought to suspend self-representation efforts through the Arctic Council when the Russian Association of Indigenous Peoples of the North was forcefully dissolved by state decree in 2012, and then re-opened with a more government-friendly leadership.

Yet another aspect deeply intertwined with both knowledge and political territory is the question of land use and resource extraction. In Greenland, the Self-Rule Act, implemented in 2009, states that the government of Greenland, Naalakkersuisut, has control of domestic affairs, including the sovereignty over natural resources, while Denmark retains control of foreign affairs and defense. As a point of comparison, contemporary discourse in Greenland emphasizes Greenlanders neither as Inuit nor as Danish subjects, partly because to be an Indigenous subject is to forego aspirations of statehood in accordance with stipulations in the ILO-169 convention. Greenland, as scholars such as Kirsten Thisted forcefully argue, is indeed a globalized, urbanized society where especially the younger generation is quite interested in foregoing talk about whether the current relationship to Denmark is colonial or post-colonial (Thisted 2013; see also Körber and Volquardsen 2014) in order to affirm a public identity, or an imagined community, that emulates expressions of national identity. Inuit of North Eastern Canada have achieved the right to self-government as an Indigenous population in the Nunavut territory (though the Inuit of Northern Québec, Labrador, and the Northwest Territories have not, nor have the approximately 15,000 Inuit who live in southern Canada), which, in the context of the long history of Canadian political discourse about the "Arctic Frontier," offers yet another version of politically, discursively, and negotiated Arctic Indigeneity and self-governance (for a helpful comparative perspective on Greenland-Nunavut self-governance movements, see Loukacheva 2007). Sherrill E. Grace (2007), moreover, argues that Canada's insistence on the "North" for formulating a national ideology is unique; no other country uses "North" as a badge of honor. Yet the Canadian government have other badges that are not nearly as honorable. Canada's "Truth and Reconciliation Report," which contained an Inuit sub-committee, was released in 2015, and the recommendations have begun, after years of neglect, to be taken on board by the government.

These examples demonstrate that the questions of territory and sovereignty so central to the concept of modernity also lie at the heart of the contested modes of politics and representation in the Arctic. Concepts of sovereignty clearly have to be expanded beyond the 1648 "Westphalian model" of territorial nation states to include other actors and agents, as scholars of European, Nordic, Canadian, and postcolonial political science have argued (Adler-Nissen and Gad 2013; Romaniuk 2013; Shadian 2014; Watt-Cloutier 2015).

HISTORIOGRAPHIES OF ARCTIC ENTANGLEMENTS

Cultural imagination plays a central role in historiography, no matter how much some historians wish to work under the guise of positivism. Arctic environments and modernities are not excluded from this process. These cultural imaginings raise a series of salient questions: What kind of cultural practices guide Arctic historiographies, who writes them, and who reads them? It is critical to acknowledge how competing and contrasting histories relate to politics, as the Arctic cannot be "imagined" as a singular nation-state "imagined community," along the lines famously formulated by Benedict Anderson (1991). Competing and contradictory histories of the same geographical place offer a form of historiography that acknowledges diversity, contestation, and a wide array of actors and agents. Seeking to replace single (national) narratives, recent historiographical theory proposes terms such as "history of entanglements," "shared history," or "*histoire croisée*" to reflect intersections, interdependencies, and competing histories (see, e.g., Werner and Zimmermann 2006; Conrad et al. 2007; Manjapra 2014; Müller et al. 2010). Environmental historians Dolly Jørgensen and Sverker Sörlin indeed echo this line of thought in their analysis of Arctic historiography:

> Is there a history of the North? Since ancient times, the answer to this question has been no. History was the narrative of human action, and where human action seemed to cease in cold and ice there could be no history …

In world historiography, the North has remained of marginal importance up until recent times, largely for the same reason: the stereotype of inaction, of little being at stake. With few exceptions, the history of northern exploration, which already a full century ago was the subject of mighty tomes and an emerging literature on economic development in the far North, served as precisely the opposite of history: a non-history of no events and the silence that preceded action. (2013: 1)

To tell the history of the Arctic from multiple perspectives, then, offers the possibility to develop new models of Arctic environmental modernities, in contradistinction to the positivist notion of a "world history," which is most often one of white, male, Christian, capitalist, colonial privilege (indeed, many of the chapters in this book delineate the conflicts and the tensions that exist at the heart of these very categories). These histories of entanglements, or shared and yet conflicting histories, foreground not a total and authoritative history, but a story that is told through the dialectical contradictions that emerge. Helga Lúthersdóttir's incisive formulation foregrounds the relevance of new historiographical approaches to the representation of Arctic environmental modernities:

Today, connotations of spaces previously "seen as literally and symbolically white" are no longer the "site of a privileged white masculinity" as the myth of no-man's land is rapidly being creolised ... Indigenous peoples now compete with neo-imperial interests in ownership of their homelands, objecting to the Anglophone concepts of "wilderness" and "landscape" dominating the discourse on the Nordic regions because such "approaches erode the appreciation of distinctively Northern and Indigenous aspects of land and life." (2015: 325)

What is important here is to recognize the multiple modernities of the Arctic; this is not simply a mode of historiographical inquiry, but recognition of the space's diversity, environmental distinctiveness, range of inhabitants, and histories.

The Book's Scope and Purpose

Arctic Environmental Modernities addresses the question as to why the Arctic plays such a key role in the cultural imagination of the future of the planet and its political and ecological systems. The book does not look at this development as a role that emerged *ex nihilo*, but instead as the culmination of over 300 years of Arctic history and the role that the emergence of environmental modernities has played in this development.

In his chapter "The Disappearing Arctic? Scientific Narrative, Environmental Crisis, and the Ghosts of Colonial History," Andrew Stuhl charts various trajectories of Arctic historiographies by addressing how 'climate change tourism' is not in any way new but instead continues a tradition of Arctic scientists concerned with the decay of environments and cultures. Stuhl examines the contradiction between the Arctic being "frozen in time" and yet being threatened, and how this relates to colonialism. He traces this contradiction through cultural ethnography, focusing on anthropologist Vilhjalmur Stefansson. Stuhl shows how Stefansson engaged in exactly the kind of environmental intervention that he warned against. The Arctic thus needs to be shifted from the object of Euro-American and Asian expansion and representation to a homeland that has been inhabited for centuries.

Synnøve Marie Vik's chapter "Petro-images of the Arctic and Statoil's Visual Imaginary" explores the visual propaganda accompanying resource extraction in the Arctic by way of the example of the Norwegian oil drilling and extraction company Statoil. Vik examines the new imagery of Arctic petro-cultures and foregrounds how the corporation's visual rhetoric trivializes the environmental impact of production sites. The most recent phase of Arctic modernization and industrialization alters its landscapes and its representations, whilst pointing to a continuity of Western capitalist imaginations of conquest and mastery of Arctic nature. Vik demonstrates the potential of visual culture approaches to a critical study of the Arctic in the Anthropocene and to the environmental humanities.

Following on from Vik's contribution, Torill Nyseth's chapter "Arctic Urbanization: Modernity Without Cities" examines another aspect of Arctic environmental modernities: urbanization. Nyseth explores current processes of urbanization in the Arctic, with a focus on northern Scandinavia, and argues that theories of urbanism have to be modified and diversified to encompass Arctic cities, most of which are far smaller in scale and population than the metropolises further to the south. According to Nyseth, Arctic urbanism today is produced within the context of a new phase of industrialization related to extractive industries and a changing geopolitical environment. The chapter provides an overview of the specificities of Scandinavian Arctic urbanity with a particular focus on the cities' multiculturalism and the characteristic proximity of culture and nature, linking Arctic urbanization and modernization inextricably with the utilization of the natural resources in the cities' environments.

Examining another aspect of Arctic modernities related to urbanization, Kristian Hvitfeldt Nielsen addresses Arctic industrialization and technopolitics in his chapter "Cod Society: The Technopolitics of Modern Greenland." Nielsen examines the consequences of the Danish Greenland Commission's grand modernization scheme for the decades after World War II, emphasizing the links established between the extraction of natural resources, new technologies, and social change in the Commission's report of 1950. The chapter focuses on two interwoven contexts: the introduction of industrial cod fishing, and the incipient urbanization and concentration in the Arctic. Drawing on the concept of technopolitics, Nielsen offers an understanding of the modernization of Greenland as a shift from one technopolitical configuration to another: the emerging "cod society" employed certain kinds of technology to restructure the postcolonial relationships between Greenland, Denmark, and the rest of the world, to introduce a modern welfare state, and to reconfigure the relation of humans and environment.

Arctic aesthetics also play a central, if fraught, role in the conception of Arctic environmental modernities. In their chapter "Re-reading Knut Hamsun in Collaboration with Place in Lule Sámi Nordlándda," Kikki Jernsletten and Troy Storfjell offer a critical and ground-breaking approach to Norwegian Nobel Prize laureate Knut Hamsun, focusing on his 1917 novel *Growth of the Soil* (*Markens grøde*). By developing a collaborative, place-based methodology grounded in Sámi cultural practices, they engage with the region Hamsun grew up in, its nature and its people, and with the emerging field of Indigenous methodologies, more specifically within the Lule Sámi context. They conclude that Hamsun's alleged modernistic style was partially, and more substantially than has been acknowledged, due to the Sámi presence in his home region. Their contribution thus not only rewrites Norwegian literary history, but offers a complementary understanding of Arctic environmental modernities.

A quite different kind of aesthetic, in this case an enviro-religious approach to the Arctic, is on display in the works of Fridtjof Nansen. In the chapter "The Polar Hero's Progress: Fridtjof Nansen, Spirituality, and Environmental History," Mark Safstrom examines the autobiographical writings of this Norwegian national hero, arguing that *Farthest North* (*Fram over polhavet*, 1897) combines polar expedition rhetoric characterized by rationality, athleticism, and male agency with religious ascetic traditions. Safstrom argues that Nansen offers a complementary model for writing Arctic environmental history in the age of nationalism, colonialism, and the emergence of industrial modernity by bringing to the fore spiritual components as constitutive of exploration. These components of Western religiosity emphasize the feminine, passive, and mythical. Safstrom juxtaposes Nansen's perspectives with Arne Naess's twentieth-century eco-philosophy. Naess's "deep ecology" emphasizes a persistent anthropocentrism that continues to influence environmental historiography about the Arctic and the Anthropocene. By synthesizing these two perspectives, Safstrom offers a new model for understanding polar exploration.

While Nansen's work can be seen as one aspect of nation building through Arctic environmental modernities, Dag Avango and Peder Roberts offer another, quite distinct, one. In their chapter "Heritage, Conservation, and the Geopolitics of Svalbard: Writing the History of Arctic Environments," the authors examine the heritage and conservation practices in Svalbard and the Spitsbergen archipelago from the nineteenth century to today. They deploy the concept of critical geopolitics to demonstrate how Arctic localities have been used as instruments of Norwegian nation building, past and present. The writing of environmental history is never separate from politics or socially and culturally inscribed power structures by which spaces construed as primordial and pristine landscapes become privileged sites for the construction of Arctic ideologies. The establishment of national parks, the construction of mining heritage sites, or the undertaking of settlement archaeology all serve such purposes, by which Svalbard becomes an icon of the culturally and politically motivated definitions of "wilderness."

Lill-Ann Körber offers a challenging and complex model for understanding the roles of the environment, climate change, and the urban in contemporary representations of global warming and maritime pollution in Greenland through Greenlandic documentary cinema and other arts. In her chapter "Toxic Blubber and Seal Skin Bikinis, or: How Green Is Greenland? Ecology in Contemporary Film and Art," Körber offers a nuanced examination of global and local ecological discourses and ecocriticism in the anthropogenic Arctic by focusing on Greenlandic artists and their agency. The films and art works analyzed by Körber portray Greenlanders as global citizens who self-consciously weigh the hazards of resource extraction and global warming against the desire for postcolonial national self-determination, challenging the potentially simplistic and patronizing set-up of "Save the Arctic" campaigns.

As Körber ably demonstrates, the cinema has functioned as a means to negotiate Arctic modernity's various identities. This negotiation is not only in play in Greenland. For instance, in her chapter "The Negative Space in the National Imagination: Russia and the Arctic," Lilya Kaganovsky delineates the ways the Arctic and Siberia are conceptualized in Russian and Soviet feature and documentary films, demonstrating that these cinematic imaginaries are not static, but constantly reconfigured through various paradigms, each set erasing or reconceiving the historical imaginary that came before. Kaganovsky examines the constitution of the Arctic in pre-revolutionary Russia, the need for Soviet expansion to unite the state, and the notion of Stalinist exploration so as to reveal how the Arctic became a tool in the creation of Soviet and Russian imaginaries through the cinema. To do so, Kaganovsky considers Vertov's *One Sixth of the World* (1926), Erofeev's *Beyond the Arctic Circle* (1927), the Vasiliev Brothers' *Heroic Deed Among the Ice* (1928), and Marina Goldovskaya's documentary on the GULAG *Solovki Power* (1988).

Moving images have also been mobilized for Arctic environmental activism. In the chapter "Invisible Landscapes: Extreme Oil and the Arctic in Experimental Film and Activist Art Practices," Lisa E. Bloom examines the global interconnectedness of the Arctic and the role of climate change and resource extraction in contemporary environmental art. Bloom analyzes the effects of replacing the "natural" with an "industrial sublime" in Ursula Biemann's *Deep Weather* (2013). She then considers the application of modes of intimacy deployed by the use of opera in Brenda Longfellow's *Dead Ducks* (2012) and charts the use of amateur aesthetics and popular culture in *The Yes Men: But It's Not that Polar Bear Thing* (2013). These experimental and activist films and videos present images of climate change and environmentalism that are not apocalyptic or sentimental; moreover, they go beyond the spectacular icons of climate change, such as calving and melting glaciers or anthropomorphized melancholic polar bears.

As moving images play a central role in the global representation of Arctic modernities, one cannot be limited solely to aesthetic and political works. Indeed, one of the key aspects of twenty-first-century image-making is advertising and its relationship to tourism, and branding. To this end, in the chapter "Icelandic Futures: Arctic Dreams and Geographies of Crisis," Ann-Sofie Nielsen Gremaud examines how, in the wake of the economic crisis in 2008, Iceland's efforts to brand itself as a modern Arctic nation state intensified. She examines how public discourse markets Iceland's clean and pristine nature and ample natural resources as a vehicle for becoming a global geopolitical player for dominance in the North. Government rhetoric conveys a consistent "Arctic optimism" on behalf of officials, as part of Iceland's attempt to leave the crisis behind and to articulate a future conceived as one of environmental cleanliness, purity, and efficiency. This "Arctic-as-utopia" discourse is often criticized in twenty-first-century Icelandic art. Art foregrounding Icelandic nature, landscapes, and environments thereby becomes an important counter-narrative to official rhetoric and a space where conflicting approaches to natural resources can be negotiated.

Eva-Maria Svensson examines the Arctic Council, a body of inter-governmental and transnational cooperation founded in 1996, from the perspective of feminist governance. In the chapter "Feminist and Environmentalist Public Governance in the Arctic," she evaluates the work of the Arctic Council with respect to both gender equality and ecological policy. One of her key findings is that Arctic residents, those directly affected by the consequences of global and local politics, are of secondary concern to Arctic policy making, and this marginalization is clouded by rhetoric that obscures this fact.

In the chapter "The Greenlandic Reconciliation Commission: Ethno-nationalism, Arctic Resources, and Post-Colonial Identity," Kirsten Thisted addresses the Greenlandic Reconciliation Commission established in 2014. Examining the different agendas and political positions that have shaped the debate around reconciliation, Thisted shows how the political processes that led to the Act on Greenland Self-Government in 2009 ran parallel to the UN negotiations on the rights of Indigenous peoples, while the term "Indigenous peoples" is not mentioned anywhere in the act. She argues that the Greenland public discourse of indigeneity is currently being transformed from a language of resistance to a language of independent governance, providing a model of global significance in terms of post-colonial identity and ethnonationalism. The Reconciliation Commission negotiations furthermore address Greenland's interest in establishing itself as an Arctic resource extraction economy while recognizing its colonial and postcolonial dependency on Denmark.

Returning to the contested definitions of the Arctic outlined in the book's introduction, Nina Wormbs and Sverker Sörlin examine Arctic policy making in a different, but equally challenging, light. In the chapter "Arctic Futures: Agency and Assessing Assessment," they argue that no other region of the world has as many scientific assessments per capita as the Arctic. This extensive assessment industry has a long history and is continuing to influence policy and politics in the region, especially with respect to questions of sustainability, resilience, and resource extraction in an era of anthropogenic climate change. Reports such as the *Arctic Monitoring and Assessment Programme* (AMAP 1997) and the *Arctic Resilience Interim Report* (2013) generate a limited view of environmental and societal "drivers" in the Arctic region, often with little concern for social and cultural complexity. They analyze the theoretical and methodological paradigms that have informed Arctic assessments in the past and the predictions for the future these reports have generated.

CONCLUSION

A guiding principle for *Arctic Environmental Modernities* is that "the Arctic" has never "existed" as a homogeneous totality, which makes it all the more contradictory that it is being fought over in such a way today. When we argue that "the Arctic" has never existed, what we are postulating is that the totalized image of the Arctic is a construct of Western Enlightenment thought, of explorers and artists, scientists and politicians, and one that is continuously promulgated today, though under the guise of the region being "open for business." Few of these conceptions originate from the inhabitants of the region. They have been, and continue to be, imposed from the outside. For these reasons, we hope that the picture that emerges of the Arctic in this collection is one that is contradictory, dialectical, and incomplete. The chapters in this collection do not in any way uphold this totalized view of the Arctic; indeed, we want to strip away this view and offer a fragmentary and dialogical account, positioning the Arctic as a site of meaning that is widely contested, continuously negotiated, reimagined, and elided. The authors of *Arctic Environmental Modernities* are therefore interested in challenging two dominant strands of Arctic research over the last decades: (i) a policy-driven governmental and geopolitical, instrumentalist approach and (ii) the deductive model of the natural sciences. Both these approaches elide the messy complexities of representational and cultural history and their implications for both the contemporary moment and the region's imagined future.

WORK CITED

Adler-Nissen, Rebecca, and Ulrik Pram Gad, eds. 2013. *European integration and postcolonial sovereignty games.* New York/London: Routledge.

AMAP (Arctic Monitoring and Assessment Programme). 1997. www.amap.no. Accessed 20 Sept 2016.

Anderson, Benedict. 1991. *Imagined communities: Reflections on the origin and spread of nationalism*. London: Verso.

ARR. 2013. *Arctic resilience: Interim report 2013*. Stockholm: Stockholm Environment Institute and Stockholm Resilience Centre.

Berman, Marshall. 1988. *All that's solid melts into air: The experience of modernity*. London: Penguin.

Bhabha, Homi. 1996. Unsatisfied: Notes on vernacular cosmopolitanism. In *Text and narration: Cross-disciplinary essays on cultural and national identities*, eds. Laura García-Moreno and Peter C. Pfeiffer, 191–207. Columbia: Camden.

Bloom, Lisa E. 1993. *Gender on ice: American ideologies of polar expeditions*. Minneapolis: University of Minnesota Press.

Bravo, Michael, and Nicola Triscott, eds. 2011. *Arctic geopolitics and autonomy*. Berlin: Hatje Cantz.

Chernilo, Daniel. 2016. The question of the human in the anthropocene debate. *European Journal of Social Theory*. Published online 2 June 2016.

Conrad, Sebastian, Andreas Eckert, and Ulrike Freitag, eds. 2007. *Globalgeschichte. Theorien, Themen, Ansätze*. Frankfurt: Campus.

Dodds, Paul, and Mark Nuttall. 2015. *The scramble for the poles: The geopolitics of the Arctic and Antarctic*. London: Polity.

Eisenstadt, Shmuel N. 2000. Multiple modernities. *Daedalus* 129(1): 1–29.

Emmerson, Charles. 2010. *The future history of the Arctic*. New York: Public Affairs.

Forte, Maximilian C., ed. 2010. *Indigenous cosmopolitans: Transnational and transcultural indigeneity in the twenty-first century*. New York: Peter Lang.

Gaonkar, Dilip Parameshwar, ed. 2001. *Alternative modernities*. Durham/London: Duke University Press.

Grace, Sherrill E. 2007. *Canada and the idea of North*. Montreal: McGill-Queen's University Press.

Harrison, Roman, and Gordon Hodgson. 1987. Forty years of *Arctic. Arctic: The Journal of the Arctic Institute of North America* 40(4): 321–345.

Hill, Jenn. 2008. *White horizon: The Arctic in the nineteenth-century British imagination*. Albany: SUNY Press.

Hobsbawm, Eric. 1962. *The age of revolution: Europe 1789–1848*. London: Abacus.

ICC (Inuit Circumpolar Council Canada). http://www.inuitcircumpolar.com/icc-international.htm. Accessed 20 Sept 2016.

Ingold, Timothy. 2013. Le Nord est partout. *Entropia* 15: 37–48.

Jørgensen, Dolly, and Sverker Sörlin. 2013. Making the action visible: Making environments in Northern landscapes. In *Northscapes: History, technology, and the making of Northern environments*, eds. Dolly Jørgensen and Sverker Sörlin, 1–13. Vancouver: University of British Columbia Press.

Keskitalo, Carina. 2004. *Negotiating the Arctic: The construction of an international region*. New York/London: Routledge.

————. 2009. 'The North' – Is there such a thing? Deconstructing/contesting Northern and Arctic discourse. In *Cold matters: Cultural perceptions of snow, ice and cold*, eds. Heidi Hansson and Cathrine Norberg, 23–39. Umeå: Umeå University.

Kjeldaas, Sigfrid and Anka Ryall. 2015. Arctic modernities. *Nordlit* 35(2): ii–iii.

Körber, Lill-Ann and Ebbe Volquardsen, eds. 2014. *The postcolonial North Atlantic: Iceland, Greenland and the Faroe Islands.* Berliner Beiträge zur Skandinavistik 20. Berlin: Nordeuropa-Institut der Humboldt-Universität zu Berlin.

Kuokkanen, Rauna. 2009. Achievements of indigenous self-determination: The case of the Sámi parliaments in Finland and Norway. In *Indigenous diplomacies*, ed. J. Marshall Beier, 97–114. New York: Palgrave.

————. 2011. Self-determination and indigenous women — 'Whose voice is it we hear in the Sámi parliament?'. *International Journal of Minority and Group Rights* 18(1): 39–62.

Loukacheva, Natalia. 2007. *The Arctic promise: legal and political autonomy of Greenland and Nunavut.* Toronto: University of Toronto Press.

Lúthersdóttir, Helga. 2015. Transcending the sublime: Arctic creolisation in the works of Isaac Julien and John Akomfrah. In *Films on ice: Cinemas of the Arctic*, eds. Scott MacKenzie and Anna Westerståhl Stenport, 328–337. Edinburgh: Edinburgh University Press.

MacKenzie, Scott, and Anna Westerståhl Stenport. 2013. All that's frozen melts into air: Arctic cinemas at the end of the world. *Public: Art/Culture/Ideas* 48: 81–91.

————, eds. 2015. *Films on ice: Cinemas of the Arctic.* Edinburgh: Edinburgh University Press.

Malm, Andreas, and Alf Hornborg. 2014. The geology of mankind? A critique of the anthropocene narrative. *The Anthropocene Review* 1(1): 62–69.

Manjapra, Kris. 2014. *Age of entanglement: German and Indian intellectuals across Empire.* Cambridge, MA: Harvard University Press.

McCannon, John. 2012. *A history of the Arctic: Nature, exploration, exploitation.* London: Reaktion.

McGhee, Robert. 2007. *The last imaginary place: A human history of the Arctic world.* Chicago: University of Chicago Press.

Müller, Leos, Göran Rydén, and Holger Weiss. 2010. *Global historia från periferin: Norden 1600–1850.* Lund: Studentlitteratur.

Robin, Libby. 2013. Histories for changing times: Entering the anthropocene. *Australian Historical Studies* 44(3): 329–340.

Romaniuk, Scott Nicholas. 2013. *Global Arctic: Sovereignty and the future of the North.* Highclere Berkshire: Berkshire University Press.

Ryall, Anka, Johan Schimanski, and Henning Howlid Wærp, eds. 2010. *Arctic discourses.* Newcastle upon Tyne: Cambridge Scholars Publishing.

Said, Edward. 1994. *Culture and imperialism*. New York: Vintage.

Shadian, Jessica. 2014. *The politics of Arctic sovereignty: Oil, ice and Inuit governance*. London/New York: Routledge.

Slezkine, Yuri. 1994. *Arctic mirrors: Russia and the small peoples of the North*. Ithaca: Cornell University Press.

The Anthropocene. www.theanthropocene.info. Accessed 20 Sept 2016.

Thisted, Kirsten. 2013. Discourses of indigeneity: Branding Greenland in the age of self-government and climate change. In *Science, geopolitics and culture in the polar region: Norden beyond borders*, ed. Sverker Sörlin, 227–258. Farnham: Ashgate.

Tuhiwai Smith, Linda. 1999. *Decolonizing methodologies: Research and indigenous peoples*. New York: Zed Books.

Watt-Cloutier, Sheila. 2015. *The right to be cold: One woman's story of protecting her culture, the Arctic and the whole planet*. Toronto: Penguin.

Werner, Michael, and Bénédicte Zimmermann. 2006. Beyond comparison. *Histoire croisée* and the challenge of reflexivity. *History and Theory* 45(1): 30–50.

CHAPTER 2

The Disappearing Arctic? Scientific Narrative, Environmental Crisis, and the Ghosts of Colonial History

Andrew Stuhl

The Arctic, we are told, is rapidly disappearing. In the summer of 2007, the extent of sea ice in the circumpolar basin plummeted to an all-time low (Revkin 2007). The popular press pounced on the shrinking ice, transforming it into the poster child of global climate change (Christensen et al. 2013, 3–9). Over the next five years, as the annual measurement of the sea ice minimum in the Arctic became a regular media event, the news only got worse. The 2008 number fell below 2007's, and new, lower records were set every year until 2012 (National Snow and Ice Data Center 2012). Subsequently, the view of the north from a satellite—enhanced with bold colors illuminating the gap between shoreline and icepack—has become an icon of a "disappearing" Arctic (Goldenberg 2013; Francis and Hunter 2006, 509–510).

The remote sensing data can hardly be denied, not to mention the observations of these conditions from people living along the northern rim. Yet, there is something lurking in the words supporting these power-

A. Stuhl (✉)
Bucknell University, 955 W. Market St, Lewisburg, PA, USA

© The Author(s) 2017
L.-A. Körber et al. (eds.), *Arctic Environmental Modernities*,
DOI 10.1007/978-3-319-39116-8_2

ful images. Ice is melting. But what does it mean to say it is *disappearing*? While such descriptions raise awareness of Arctic issues for outsiders, they unleash a curious social power within the region. As scholars have shown, the issue of sea ice reveals how science and narrative work together in framing environmental crises in the Arctic—and the most appropriate responses to them (Christensen 2013; Wormbs 2013; Huntington 2013; Ryall et al. 2008, x–xxi). As the Arctic thaw has revealed previously inaccessible natural resources and transportation routes, reports about a warming north have helped turn the region into a hive of economic prospecting and governmental capacity building (Avango et al. 2013, 431–446; Martin 2014). Should that seem a heretical claim, or one insensitive to the complexities of communicating science, consider the view of some Arctic residents. According to Inuit leaders Duane Smith and Mary Simon, the reaction to sea ice retreat by developers and federal governments has marginalized Inuit participation in circumpolar governance, even within the intergovernmental forums long dedicated to the advancement of indigenous interests—like the Arctic Council and the Kelowna Accord (Simon 2008; Rynor 2011; Exner-Pirot 2012; Hossain 2013). Importantly, scientific narratives about vanishing ice in the Arctic effect this marginalization because they conjure ghosts from the region's colonial past.

   In this chapter, I will build on critical perspectives of the media of Arctic climate change by locating and analyzing the historical predecessor of today's "disappearing" Arctic. Scientists in the early twentieth century deployed similar language to describe the first great environmental crisis in the far north. Between the 1880s and the 1920s, populations of caribou and Inuit along the north-western coast of North America dropped sharply. The culprit? The global whaling industry. In search of baleen, a precious commodity used in the fashion industry, whalers emptied the Beaufort Sea of bowheads. Outposts and trading villages cropped up on the northernmost shores of the continent, setting the stage for the first sustained contact between visitors, the diseases they brought with them, and Inuit (Bockstoce 1986). The results were shocking. At the mouth of the Mackenzie River, Inuit communities that might have hosted more than 2500 individuals in the 1860s dwindled to 259 people by 1905 (Arnold et al. 2011, 81). And caribou in the area, the primary source of sustenance for locals and the thousands of whalers joining them, died off just as quickly (Beregud 1974).

Scientists in the early 1900s agreed that these events constituted the end of the Arctic itself. Of his own work recording Inuit relationships with the northern environment in the 1910s, anthropologist Diamond Jenness wrote, "were we the harbingers of a brighter dawn, or only messengers of ill-omen, portending disaster?" (Jenness 1928, 247). The answer to that question came just a few years later. Jenness and other anthropologists converted their narratives into governmental interventions, building colonialist natural resource management programs out of the fear and scientific authority of their "disappearing" Arctic rhetoric.

Drawing from scientists' published works, as well as Inuit versions of the same events, I examine the "disappearing" Arctic of the early 1900s. I begin by establishing the cultural moment in which the Arctic emerged as both a target of social concern and a critical field site for the development of science. This context sets up a summary of Inuit perspectives on the crisis of the early 1900s and a detailed investigation of two works of popular science that became authoritative accounts of this issue as it unfolded: Vilhjálmur Stefánsson's *My Life with the Eskimo* (1913) and Diamond Jenness's *People of the Twilight* (1928). My goal is not to suggest a direct analog to today, whereby the past and present ought to be held up for direct comparison. Rather, the point is to shed light on the historical force of our language, especially that deployed in scientific media to refer to Arctic nature. As a history of science clearly shows, these words contain much more historical baggage than is suggested by their proliferation across the headlines.

DARWIN'S SHADOW: SCIENTIFIC WORLDVIEWS AND SOCIAL CONCERNS CIRCA 1900

To emphasize representations of environmental crises in the Arctic as functions of scientific practice, it is helpful to begin not with a description of the material changes taking place in the Arctic in the early 1900s, but with a characterization of the culture of science and social concerns at that time. Such an approach establishes knowledge-making practices as a medium through which Euro-Americans have understood the non-human world.

During the late nineteenth and early twentieth centuries, popular and scientific audiences in North America became preoccupied with the decay of nature and society (Gruber 1970, 1289–1291). This fear expressed itself in many forms, including the Eugenics Crusade, rural modernization initiatives, anti-immigration and sanitation projects in cities, and the conservation and wilderness movements (Paul 1995; Warren 2003, 180–241; Lovett 2007). While a diverse array of cultural values underpinned these activities—from ideals of racial purity to the guilt of exploiting nature for human gain—a single scientific worldview was common to them all: evolutionary theory (Paul 1995). Thus, in order to understand why scientists would be convinced the Arctic was on the verge of disappearing in the early 1900s—and the currency of that turn of phrase today—we must pick apart some of the intricacies of Darwin's theory.

This theory understood the history and development of all life as fueled by a competition for scarce resources needed for sustenance. As time rolled on, so did natural selection: those species unable to secure their own livelihoods would die out, leaving only the fittest to survive (Worster 1994, 145–169). Through the lens of Darwinism, then, one could resolve the spatial distribution of plants and animals on earth into identifiable regions and associated underlying processes—a broad, interdisciplinary pursuit referred to as "biogeography" (Cox and Moore 2005, 15–43). Oak savannas exist here and not there, for instance, because of the geology of this soil and because the biological advantage of broad leaves gradually eliminated thin-leaved species from this area. In these ways, evolutionary theory allowed a range of sciences to mature at the close of the 1800s. Geology, biology, paleontology, and climatology were all enrolled in the quest to understand the origins and mysteries of nature (Bowler 1995; Rainger 2004).

Because it arranged the world into units—and afforded nation states more efficient control over resource development within those units—biogeography has been understood as a colonial tool and a salvation from social anxieties about disorder in the 1800s (Browne 1983). Such saving properties are even more apparent in the application of evolution to the study of people. Just as evolution removed ill-adapted traits and behaviors from plants and animals, so too would the process eventually create the most complex human communities with the most advantageous relationships with the non-human world. Museum-based scientists in Europe and the United States operated from this assumption to lay the foundations for modern anthropology, archaeology, geography, and sociology (Hinsely 1981, 7–10; Smith 2003). They classified the world's human groups according to three points on a spectrum: savagery, barbarism, and enlightenment (civilization) (Hinsely 1981, 88). These stages were meant to parallel the general development of nature from wilderness to farm-land to industrial city. Because they had spread so widely over space and time, the beliefs, technologies, and social structures held by Europeans and North Americans were understood to be well adapted and thus supe-rior. This scientific view, while racist in the clarity of hindsight, formed one strand of a "Social Darwinism" wildly popular among scientists and urban residents in the late 1800s (Kevles 1985; Stocking 1968, 113–132).

As the Darwinian revolution of the 1800s became common sense in the early 1900s, the English language absorbed ways of talking about people and nature that bespoke evolution and the cultural moment of the theo-ry's birth. Plants that showed an ability to expand their geographical range across landforms and climates became known as "pioneers," "invaders," and "colonizers" (Nyhart 2010, 49–58; Bowler 1995). Associations of plants and animals that scientists identified as characteristic of a particular latitude, temperature range, or altitude became "provinces," a new geo-graphical term and space that muddled the lines between science, nature, and colonialism (Browne 1983). Most importantly, "lower" or "elemen-tary" forms of plants, animals, or humans—Neanderthals, for example—would be eliminated by higher orders of life, as the fossil record made clear (Bowler 1995). To study evolution in the early 1900s, then, one had to confront the process of extinction and articulate the moment of disappearance.

Evolutionary Science in the Arctic; the Arctic in Evolutionary Science

From the start, the Arctic generated special scientific interest in compre-hending the history of nature, the evolution of mankind, and the meaning of progress. In the United States, scientists at the Smithsonian Institution based classification methods and ethnographic theories on collections streaming into Washington from Arctic and polar exploration (Hinsely 1981; Fitzhugh 2009). Because museum-based anthropologists had at their disposal an array of Inuit artifacts from across the circumpolar basin, they viewed these materials as an unparalleled dataset for organizing human society according to its historical development. From his position as curator, Otis Tufton Mason noted that, despite their geographic differences, Inuit communities all deployed a "throwing stick" as a weapon. Based on this observation, Mason developed a style of analysis called the "ethnographic unit" which exhibited and categorized Inuit culture according to its reliance on the fauna and flora of a given landscape. Mason, like biogeographers, presumed that the Arctic environment was internally homogeneous. From this point, scientists rendered nature a determinant variable in the production of culture (Hinsely 1981; Darnell 1998; Bravo 2002).

Indeed, assumptions about the influence of the Arctic environment on Inuit inspired revolutions in anthropological science, even as they perpetuated evolutionary theory's emphasis on extinction. Responding to Mason's conclusions, Franz Boas—long recognized as a father of American anthropology—criticized the "ethnographic unit" approach for misconstruing the human relationship with the non-human world (Darnell 1998, 2–7; Stocking 1987, 284–292). It was probable, Boas claimed, that "unlike causes produce like effects" (Boas 1887, 485). He demanded a careful study of how the human mind reacts to environmental pressures in different situations before postulating why Inuit constructed throwing sticks where they did (Hinsely 1981). Indeed, Boas's own field research in the Arctic—on which his seminal *The Central Eskimo* (1988) was based—suggested new empirical and theoretical bases for anthropology (Cole 1983, 13–17). For Boas, classification did not equate to explanation. Anthropologists, he charged, ought to rest their discipline not solely on objects, but also on the investigation of language, beliefs, folklore, and, most of all, the historical development of these phenomena (Stocking 1992a, 62). Yet, even these Boasian methods led the discipline to more relativistic definitions of culture—and thus away from the racism implicit in the savage-barbarian-civilization model—they preserved aspects of evolutionary thinking at the core of anthropological study.

In interpreting cultures, their relationships with local environments, and their similarities and differences across time, there was no more important element for Boas and his students than the moment of "contact" (Stocking 1992b, 119–161). Such engagement with the Arctic encouraged scholars to set Inuit apart from the planet and thus position themselves as observers of the initial "encounter." If such scientific framing allowed anthropologists to study Inuit in more depth, it also fostered a narrow view of Arctic history. Franz Boas, for example, excised data from *The Central Eskimo* (1888), which would have established Inuit of the Eastern Arctic as intimately connected with the capitalist economies of the North Atlantic (Searles 2006, 92–94). As scientific travelers followed Boas to the Arctic in the early 1900s, they continued to interpret its climate as harsh and its landscape as barren, almost frozen in time, and thus not yet folded into the world system (McCannon 2012, 125–235; Fienup-Riordan 2003, 27–40). These ideas persisted despite observations that Inuit communities regularly procured food resources, clothing, and shelter from a variety of northern ecosystems and trade networks (Bockstoce 2009; Inuvialuit Cultural Resource Center 2016). Such discursive devices continue to undergird science and ideas about the Arctic. Scientists regularly position the northern regions as beyond the edge of the modern experience and thus relatively intact as a field site for studying what they call the most basic, most historical, and most unadulterated forms of life (Stuhl 2013).

This culture of science in the early 1900s helps contextualize the crisis of declining caribou and Inuit populations, though I have yet to detail fully those events. In the eyes of scientists, the Arctic was a relic of primitive Earth, a sort of living fossil. As anthropologist Vilhjalmur Stefansson wrote in 1913, "[Inuit] existence on the same continent with our populous cities was an anachronism of ten thousand years in intelligence and material development" (Stefansson 1913, 2–3). Following Boasian and evolutionary worldviews, the North's wilderness and untouched native cultures were surely to be eliminated as civilization encroached and evolution ran its course. Thus, as reports of these very kinds of changes came forth in the first decade of the 1900s, the scientific communities of Canada and the United States leapt into action.

MY LIFE WITH THE ESKIMO (1913) AND PEOPLE OF THE TWILIGHT (1928)

The effects of whalers on the Arctic first emerged from reports of missionaries and northern police, who concentrated on the introduction of alcohol and prostitution aboard overwintering vessels (Bockstoce 1986, 191–203). Soon, however, scientists turned the events from a localized instance of unsavory sociality into an exemplar of the necessary, but unfortunate consequences of human progress on all life. Such portrayals helped garner financial and institutional support for a series of governmental interventions meant to isolate Inuit from modernity and protect big-game species. In many cases, these interventions served only to further the project of colonialism in the western Arctic (Levere 1993; Sandlos 2007; Piper and Sandlos 2007).

The paper trail of this history can be found scattered across many places—in the annals of scientific journals; the archives of newspapers like *The New York Times*, which covered Arctic affairs regularly; in the files of the Canadian federal departments and the American Museum of Natural History that financed the ventures; and in the correspondence of Arctic scientists with their supervisors, colleagues, families, and confidants. While I draw from these sources to trace a scientific understanding of environmental crisis in the Arctic during the early 1900s, I focus my analysis on two works of popular anthropological science—Vilhjalmur Stefansson's *My Life with the Eskimo* (1913) and Diamond Jenness's *The People of the Twilight* (1928). Stefansson and Jenness were leading anthropologists of their time and became trusted consultants to the arms of the US and Canadian governments interested in Arctic matters between 1910 and 1960 (Peyton and Hancock 2008; Kulchyski 1993; Stuart Jenness 2011; Palsson 2001). Their books, which launched their respective careers, were geared to general readers, unlike the specialized academic audiences they targeted with research articles. As such, *My Life with the Eskimo* and *The People of the Twilight* constitute suitable historical counterparts to modern day scientific media, as they seek to convey to the public at large the scientific basis of environmental crisis.

As was common in anthropological writing at the time, both books cover a range of topics—from exploration to natural history to archaeology to measurements of Inuit bodies (Clifford 1988). While the breadth of coverage provided readers with a holistic picture of the north, both Stefansson and Jenness did not miss the opportunity to repeat a consistent theme across their work: the Arctic, as Euro-American society knew it, was threatened. In *People of the Twilight*, the reader glimpses this message on the very first page. Contact with modern civilization, the author writes, "will upset the whole system of [Inuit] life and community, and they must sink" (Jenness 1928, xi). Stefansson similarly underscored the rapidity and scale of change unfolding at the top of the world. "The time from 1889 to 1906 is but a few years," he noted, yet the damage wrought on Inuit in that time had been greater than anything experienced in northern Canada "in a hundred years" (Stefansson 1913, 40). Early twentieth-century readers would have read these descriptions by trained anthropologists as evidence of the certain elimination of Inuit culture from the face of the earth, yet another notch in the belt of evolution.

Stefansson and Jenness characterized Arctic change by observing the ways Inuit and the physical environment responded to the arrival of whalers to the region. Stefansson noted that whalers intensified pressure on local caribou herds, a source of food and clothing for natives (Stefansson 1913). Jenness predicted that, in less than a generation, "there will be practically no caribou at all" in the far north (Jenness 1921, 550). A declining caribou population, in turn, forced the migration of Inuit from their traditional hunting grounds. More damaging, in scientific opinion, was the subsequent dependence Inuit had on foreign goods supplied by whalers, evidenced by shifts in Inuit clothing choices, house-building practices, and religious identifications (Stefansson 1909, 601; 1913, 40–1). This last point was especially important in light of evolutionary theory. As Stefansson explained, "the lower you go in the scale of human culture the more religion you find." Eskimo once had so much religion "that a man scarcely turns his hand over without the act having a religious significance" (Stefansson 1913, 38). Thus, by tracking the penetration of what they called civilization into Inuit communities on the Beaufort Sea— through caribou, cotton, canvas, or Christianity—Stefansson and Jenness simultaneously documented a "lower form" of human society and that form's disappearance.

Stefansson and Jenness distributed the blame for these changes. They rebuked the whaling industry, even as both scientists regularly contracted whaling vessels to travel and ship collections back home (Stefansson 1913, 368). Indeed, because scientists utilized the emerging transportation networks established by Inuit and whalers, they could more easily observe the regional changes to caribou populations unleashed by intensified hunting (Palsson 2003, 82–84). Jenness pointed to a specific whaler in a specific moment in time for catalyzing the erosion of traditional hunting methods and social structure among Inuit—Joseph Bernard, in 1910 (Jenness 1917, 91; 1921). Missionaries also became villains. Stefansson chastised the project of "civilizing" the Inuit, calling it genocide. To dramatize this point, he encouraged patrons of mission stations to request photos of Inuit graveyards, as the Church's legacy in the Arctic would be a dead Inuit culture (Stefansson 1909, 601; Diubaldo 1978). Importantly, anthropologists absolved Inuit from responsibility for what was transpiring around them. Anticipating the readers' familiarity with the story of the buffalo in the American West, Jenness invited his audience to consider a scientifically sympathetic view of Inuit: "how can we expect a primitive people like the Eskimos to make regulations for the preservation of their game?" (Jenness 1928, 153). Like Stefansson, he saw the rifle as a primary cause of the caribou's destruction and the contamination of Inuit culture.

Thus, once again, anthropologists indicted whalers and civilization itself for spoiling the Arctic forever (Stefansson 1913, 264).

In making these rhetorical moves, Stefansson and Jenness crafted a powerful colonialist narrative. They tapped into three tropes already well established in Euro-American scientific writing by the early 1900s. First, they presented the regulation of hunting as a hallmark of enlightened society (Jacoby 2001). Without it, Inuit were hopeless, as they exterminated the very food source on which their way of life rested. Second, and related, anthropologists implicitly recommended the Arctic could only be saved by restraining those Inuit with rifles from hunting and preventing the rest from obtaining rifles at all (Sandlos 2003, 2006). Finally, anthropologists positioned themselves as the best shepherds of this process, a move that sought to displace existing colonial agents like missionaries, police, and teachers (Wheeler 1986; Clifford 1988; Gupta and Ferguson 1997, 8–14). Sometimes, the anthropologists didn't just hint at this strategy. Stefansson minced no words in pushing for scientifically managed intervention, titling one popular treatise, "The Eskimo and Civilization: Disease and Death for the New Eskimo Tribes with Pauperization of those that Chance to Survive can be Prevented only by a Quarantine which will allow the Conditions of Civilization but Slow Entrance to their Territory" (Stefansson 1912). In these ways, the "disappearing" Arctic often furthered the very project of intervention it warned about.

INUIT PERSPECTIVES ON THE FIRST ENVIRONMENTAL CRISIS IN THE ARCTIC

Since this chapter offers a historical account of the ways scientists narrated environmental change at the turn of the twentieth century, it is important to insert an analytical distance between my reading of the past and the scientists' own work, which often forms the basis of environmental history. To achieve this distance, I now review histories written by Inuvialuit authors living in the region today, whose ancestors encountered the very fieldworkers who recorded changes in caribou populations and Inuit lifestyles. While Inuit narratives have their own biases, they nonetheless allow non-Inuit scholars to appreciate how social and environmental conditions can be understood outside of dominant scientific worldviews and Euro-American concerns. These narratives, then, allow us to make visible the social relations at stake in notions of the "disappearing" Arctic, whether in the past or present.

A recent history produced in part by Inuvialuit authors—the Inuit of Canada's western Arctic—immediately demonstrates alternative means of interpreting history and the Arctic experience. In this historical account, the coming of Europeans and American strangers, or *Tan'ngit* in the Inuvialuktun language, marks the start of a transition between a traditional period (1300 to 1800) and the era of modern land claim agreements (the 1970s to now) (Arnold et al. 2011). Felix Nuyaviak, an Inuvialuit who lived between 1892 and 1981, often told a story in which whalers were depicted as giants who "took pleasure in scaring and frightening game and people" (Arnold et al. 2011, 53). These giants were responsible for dispersing Inuit from where they lived along the Beaufort Sea coast and creating divisions among them. That whalers became the stuff of Inuvialuit legend can be further appreciated when comparing it to the place of British explorers in Inuit history. The naval officers who repeatedly sought the Northwest Passage in the first half of the 1800s, and who subsequently achieved so much fame in Euro-American accounts of the past, are not incorporated in the foundations of Inuvialuit history. According to native northerners, these visitors left "little lasting impression on the Inuvialuit" (Arnold et al. 2011, 54).

This understanding of time underscores an Inuvialuit worldview, shifting the Arctic from the edge of Euro-American expansion to a homeland with centuries of use and occupation. It also asserts the conviction that Inuit did not disappear in the early 1900s. On the contrary, the authors emphasize that past crises strengthened Inuit and their relationship to the land, rather than destroying both. Disease epidemics; the exhaustion of whale, caribou, and fur-bearing creatures; the spread of Christianity throughout Inuit communities: all of these events earn their own sections in Inuvialuit history and "help the next generation ... build the pride and confidence to continue to support the development of the Inuvialuit as a culture and society" (Arnold et al. 2011, foreword). This is a point Inuvialuit convey to scholars of Arctic studies, including archaeologists and anthropologists who have referred to historical extinctions of some Inuit communities (Cockney 2012).

Whereas anthropologists villainized the global whaling industry for introducing technologies that polluted an "uncontaminated" people, Inuvialuit describe how these technologies became essential to personal and cultural benefit. Canadian and Alaskan Inuit appreciated new hunting and trapping equipment—like the leg hold trap—as more efficient than their existing methods, which relied on blocks of ice to crush foxes (Arnold et al. 2011, 63). Relying on negotiation skills they had built up through centuries of trade with Inuit across the northern rim, Inuvialuit bartered specifically for this item in exchange relations with whalers. Similarly, the Inuvialuit recognized the advantages of sail-powered boats, which were able to navigate close to the wind, over their own *umiaks*, which could not (Arnold et al. 2011, 63). Combining the leg hold trap and the sail-powered boat, Inuit were able to expand and intensify fox trapping in the years between 1910 and 1970, creating the basis for the modern settlement geography and identity of Inuvialuit (Arnold et al. 2011). This was the same process relayed by anthropologists as a march of death forced upon Inuit by whalers and the rifle.

Indeed, the appropriation of non-Inuit culture into Inuvialuit ways of life allowed native northerners to withstand a greater threat than Euro-American technology: Euro-American diseases. Inuvialuit recorded six separate disease outbreaks among their people in the years corresponding with the coming of whalers—including epidemics of measles, mumps, smallpox, and influenza (Arnold et al. 2011). The waves of disease crashing upon Inuit society surely reflected native northerners' lack of immunity to Euro-American biological invaders (Piper and Sandlos 2007). Inuvialuit culture persisted through the disruptions that sickness inflicted on family life, but it certainly did not remain unscathed. Disease encouraged some to "adopt the ways of the Tan'ngit," including Christianity, once shamans proved incapable of curing the ailments (Arnold et al. 2011, 80). It also necessitated even closer interaction with northern police, whalers, and scientists to secure support for Inuvialuit lifestyles and patrol the activities of an increasing number of colonial agents. Given these conditions, Inuvialuit narrate the arrival of whalers and their diseases in the Arctic as intensifying the hybridization of Inuit and non-Inuit worlds, rather than one of these worlds consuming the other. In these accounts, Inuit—not anthropologists—served as brokers of culture and nature. They decided what constituted Inuit-ness and how to live on the shifting Arctic environment (Arnold et al. 2011). In short, without whalers, Inuvialuit and the Arctic could not be what they are today.

Like scientists of the early 1900s, modern Inuvialuit associate a history of illness with the decline of caribou in the western Arctic. Accounts differ, however, in the meaning ascribed to caribou numbers. Between 1890 and 1908, whalers ate more than 12,000 caribou at Herschel Island, a port just offshore of the northern Yukon slope and at the center of the Arctic whaling enterprise (Bockstoce 1980). Inuvialuit state that, as caribou became harder to find and disease cases multiplied, those Inuit who survived turned back toward the whaling trade. There, Inuvialuit replaced caribou hides, bones, and meat with American goods, like store-bought foods, canvas, and rifles (Arnold et al. 2011). Some Alaskan Inupiat followed the expanding whaling industry as it moved eastward along the Beaufort Sea coast and into the Mackenzie Delta of the Canadian Arctic. Descendants of these families remember the journeys to search for caribou and secure livelihoods through the bustling trade in the area (French 1992; Okpik 2005). Even as such memories detail tragic loss—of loved ones, of beloved places—they often stress the adaptation and resilience of Inuvialuit people in the face of environmental change (Arnold et al. 2011; Cournoyea 2012). This is a rather different narrative than one that appeals to scientific truth by professing the end of Arctic life.

THE LEGACIES OF A "DISAPPEARING" ARCTIC: CREATING GOVERNMENTAL AGENCY, CONTAINING INUIT AGENCY

While it is common for different groups of people to narrate change in contrasting ways, the legacies of this particular instance are unusual. Traces of Stefansson and Jenness's influence can be found across the north—from family lineages in the Beaufort–Delta region to the name of a high school in Hay River, Northwest Territories (Palsson 2008). Indeed, the very accounts of Inuvialuit history I've summarized here refer to Stefansson and Jenness to describe the habits, beliefs, and social structures of pre-Christian Inuit society (Arnold et al. 2011). What has been hitherto unrecognized, however, is how the trope of a "disappearing" Arctic inflected the management of Arctic natural resources in the first half of the twentieth century—and might also direct it in the first half of the twenty-first.

In an age before computers, airplanes, satellites, and the adoption of English in northern North America, the images and stories of Arctic change required incredible productive labor in order to register in political circles. Before being identified as a problem, the particular features of this crisis—declining caribou, greedy whalers, dying Inuit—had to be observed and assembled into a narrative that struck the chords of metropolitan society. We've already seen how these events transpired in relation to anthropological fieldwork, evolutionary theory, the social concerns of the early 1900s, and works of popular science. But, in order for such narratives to shape governmental action, they had to be circulated to appropriate audiences by credible experts. Because important components of this work are carried out by communications technologies today, they are often concealed from public scrutiny. As a result, the notion of an Arctic disappearing because of sea ice melt can seem as objective or inert as the satellite image undergirding it. Deepening our historical perspective on science explodes such a conclusion.

After publication, Stefansson and Jenness repeated the storylines of *My Life with the Eskimo* and *The People of the Twilight* over and again. They proclaimed the coming destruction of the Arctic—as well as their own plans for saving it—in lecture tours, newspaper editorials, and in daily conversations that escaped the pens of history (Swayze 1960, 52–94; Stuart Jenness 1991; Diubaldo 1978; Palsson 2001, 2003). This circulation of concern about Inuit and the northern environment both legitimized an increasing US and Canadian governmental presence in the Arctic and promoted the importance of anthropological science within the colonial apparatus (Peyton and Hancock 2008). In 1919, Stefansson and Jenness helped convene a formal public inquiry on the state of the Arctic. Following the testimony of these two anthropologists, as well as many other missionaries, northern police, whalers, and fur traders, the Canadian Department of the Interior decided to introduce reindeer—a species foreign to the North American Arctic—as a civilizing project and an experiment in scientifically managed industrialization for the western Arctic (Sandlos 2007; Stuhl In press). This inquiry, then, makes visible the discursive and *material* power of the "disappearing" Arctic narrative.

The logic behind the reindeer project tied together evolutionary theory, Arctic anthropology, and colonial aspirations for the tundra. Scientists hoped reindeer could replace the roles caribou had played in Arctic ecology and Inuit culture, and thus preserve both. They also hoped that by giving up hunting for herding, Inuit would climb the evolutionary ladder toward enlightenment. Meanwhile, the logistics of herding aligned neatly with Canadian dreams of developing the Arctic. Inuit herding units could be more easily contained—geographically speaking—thus clearing space for extensive mineral exploration. And, the excess meat produced through raising animals could feed the growth of the mining industry (Stuhl In press; Demuth 2013).

As witness that Inuit also recognized the conditions at the foundation of these colonial schemes—but interpreted them differently—Inuvialuit embraced the idea of reindeer. Mangilaluk, an Inuit leader living in what is now the Tuktoyaktuk region, rejected governmental proposals to enter a treaty agreement and recommended instead that Canadian officials bring reindeer to the area (Hart 2001, 14). While Inuvialuit supported the introduction, they chafed against what came with the animals. When Stefansson and Jenness consulted with the Canadian government on how to build a reindeer program, they made close supervision of Inuit a necessity, given that these "primitive hunters" lacked experience, intelligence, and prudence (Stuhl In press; Jenness 1923). Inuvialuit remember their relations with reindeer program officers as strained. In comparison with the fur trade, which they were helping elaborate at the same time, Inuvialuit found the tasks of herding monotonous and incompatible with their preferred social lives (Hart 2001; Stuhl In press). For these reasons, those in the Mackenzie Delta area avoided signing up as apprentices in the reindeer program and invested fully in the fox trapping economy (Hart 2001; Arnold et al. 2011). Stefansson and Jenness perceived this Inuit response as further evidence of their loss of culture and the continuing entrenchment of industrial society in the north (Interdepartmental Reindeer Committee 1933). They doubled-down on reindeer as the key to saving the Inuit and the Arctic, even as reindeer programs failed to recruit native herders (Stuhl In press; Hart 2001).

While there are many legacies of the reindeer program, one deserves highlighting here. That is, narratives circulated by scientists in the public realm have unique influence on the shape of governmental and Inuit agency in the Arctic. That Inuvialuit had to avoid consciously these programs in order to remain beyond the reach of colonial administrators evinces the colonial thrust of *My Life with the Eskimo* and *People of the Twilight*. Current media reports about global change in the Arctic refer specifically to ice and not Inuit culture, of course. But the colonialist momentum is still there, packed inside the language. The term describing these changes—"disappearing"—is so loaded with colonial history that its force continues to operate.

Geographer Emilie Cameron has convincingly shown that such power dynamics are at play in climate science today, specifically in the literature on vulnerability and adaptation (Cameron 2012). She argues that scientific representations of Inuit and their relations with the Arctic environment with terms as seemingly innocuous as "traditional" and "local" end up extending "colonial forms of knowledge and practice." Scientists often frame Inuit ecological knowledge as inapplicable to environmental conditions spawned by climate change. As a result, Inuit lose standing in natural resource management decisions. "Government intervention has never been neutral for northerners," Cameron concludes, "especially when it was well-meaning and designed to specifically prepare Inuit for a changing world" (Cameron 2012, 111).

Climate scientists and the journalists who communicate their research do not intend to harm Inuit by reporting on the extent of Arctic sea ice. They probably do not think their articles capable of such a feat. Yet, the very words they use to document climate change—and inspire global awareness of the issue—set the terms of popular, political, and economic engagement with the Arctic. In other contemporary Arctic issues, like the subsistence seal hunt, environmental risk assessment, and oil exploration, Inuit have brought considerable attention to the colonial politics at work in the words used to represent Arctic life (MacNeil 2014; Inuit Tapirit Kanatami 2013; Cassady 2010). As with the first environmental crises in the far north, the friction of these modern controversies revives colonial history and reveals the relationship between scientific narrative and human agency.

CONCLUSION

If the ghosts of colonialism's past can manifest in a turn of phrase, one hopes attention to our language can keep history from repeating itself. In this chapter, I've offered a history of science as a meditation on the cultural baggage of particular ways of talking about social and environmental change in the far north. Such historical perspectives are useful tools in understanding the Arctic today. History—of science, in particular—advances processes and language that respects both Inuit sovereignty and the decolonized scientific method.

Indeed, anthropology—the same field complicit in forwarding a set of oppressive tropes about the Arctic—is leading the way in decolonizing science, its practices, and its narratives (Smith 2012; Harrison 1997). The discipline has accomplished this by confronting its own past and incorporating the interrogation of anthropology's ties with colonialism in the training of its practitioners. Such a reflexive, critical, and robust understanding of science has precipitated another revolution in anthropology to rival that following Mason or Boas. Those once deemed "subjects" are now expert collaborators, and the museum has transformed from colonial archive to contact zone (Clifford 1997). This is especially true in the western Arctic, the very place where anthropologists developed the foundations of their discipline and spearheaded governmental interventions. Archaeologists now work directly with Inuvialuit and Inupiat to repatriate collections, develop museum and online exhibits, and narrate the meaning of Inuit identity in a changing world (Lyons 2007, 2009; Lyons et al. 2011).

What such partnerships should suggest for climate scientists today, then, is nothing less than a new standard for scientific practice. Building on the important protocols established for community-based environmental monitoring in the Arctic, climate science needs to become decidedly historical (Cuomo et al. 2008; Government of the Northwest Territories 2014). From fieldwork to final publication, scientists must consider what has come before them as they work alongside Arctic residents to interpret the changes they see all around them. Only when we begin to exorcise science's colonial history will the path forward finally appear.

Work Cited

Anne-Claude Martin. France wants to spearhead EU strategy on Arctic region. *Euractiv.* 2014. http://www.euractiv.com/sections/sustainable-dev/france-wants-spearhead-eu-strategy-arctic-region-303559. Accessed 8 July 2014.

Arnold, Charles, Wendy Stephenson, Bob Simpson, and Zoe Hoe, eds. 2011. *Taimani: At that time. Inuvialuit timeline visual guide.* Inuvialuit Regional Corporation.

Avango, Dag, Annika Nilsson, and Peder Roberts. 2013. Assessing arctic futures: Voices, resources and governance. *The Polar Journal* 3(2): 431–446.

Beregud, Arthur T. 1974. Decline of caribou in North America following settlement. *The Journal of Wildlife Management* 38(4): 757–770.

Boas, Franz. 1887. The occurrence of similar inventions in areas widely apart. *Science* 9(224): 485–486.

Bockstoce, John R. 1980. The consumption of caribou by Whalemen at Herschel Island, Yukon Territory, 1890 to 1908. *Arctic and Alpine Research* 12(3): 381–384.

———. 1986. *Whales, ice, and men: The history of whaling in the Western Arctic.* Seattle: University of Washington Press.

———. 2009. *Furs and frontiers in the far North: The contest among native and foreign nations for the bering strait fur trade.* New Haven: Yale University Press.

Bowler, Peter J. 1995. The geography of extinction: Biogeography and the expulsion of 'Ape Men' from human ancestry in the early twentieth century. In *Ape, man, ApeMan: Changing views since 1600,* eds. Raymond Corbey and Bert Theunissen, 185–192. Leiden: Department of Prehistory.

Bravo, Michael. 2002. Measuring Danes and Eskimos. In *Narrating the Arctic: A cultural history of nordic scientific practices,* eds. Michael Bravo and Sverker Sörlin, 235–273. Canton: Science History Publications.

Browne, Janet. 1983. *The secular ark: Studies in the history of biogeography.* New Haven: Yale University Press.

Cameron, Emilie S. 2012. Securing indigenous politics: A critique of the vulnerability and adaptation approach to the human dimensions of climate change in the Canadian Arctic. *Global Environmental Change* 22: 103–114.

Cassady, Josslyn. 2010. State calculations of cultural survival in environmental risk assessment: Consequences for Alaska natives. *Medical Anthropology Quarterly* 24(4): 451–471.

Christensen, Miyase. 2013. Arctic climate change and the media: The news story that *was.* In *Media and the politics of Arctic climate change: When the ice breaks,* eds. Miyase Christensen, Annika E. Nilsson, and Nina Wormbs, 26–51. New York: Palgrave Macmillan.

Christensen, Miyase, Annika E. Nilsson, and Nina Wormbs. 2013. Globalization, climate change and the media: An introduction. In *Media and the politics of Arctic climate change: When the ice breaks,* eds. Miyase Christensen, Annika E. Nilsson, and Nina Wormbs, 1–25. New York: Palgrave Macmillan.

Clifford, James. 1988. *The predicament of culture: Twentieth-century ethnography, literature, and art.* Cambridge, MA: Harvard University Press.

———. 1997. *Routes: Travel and translation in the late twentieth century.* Cambridge: Harvard University Press.

Cockney, Cathy. We are still here: Inuvialuit cultural revival and adaptation. Presentation before the 18th Inuit Studies Conference, October 26, 2012, Washington, DC.

Cole, Douglas. 1983. 'The value of a person lies in his herzenbildung: Franz Boas' Baffin Island Letter-Diary, 1883–1884. In *Observer's observed: Essays on ethnographic fieldwork,* ed. George Stocking, 13–52. Madison: University of Wisconsin Press.

Cournoyea, Nellie. 2012. Adaptation and resilience: The inuvialuit story. A speech before the 18th Inuit Studies Conference, Washington, DC, October 26, 2012.

Cox, C. Barry, and Peter Moore. 2005. *Biogeography: An ecological and evolutionary approach.* Malden: Blackwell Publications.

Cuomo, Chris, Wendy Eisner, and Kenneth Hinkel. 2008. Environmental change, indigenous knowledge, and subsistence on Alaska's North slope. *Scholar and Feminist Online* 7(1): 14.

Darnell, Regna. 1998. *And along came boas: Continuity and revolution in Americanist anthropology.* Philadelphia: J. Benjamins.

Demuth, Bathsheba. 2013. More things on heaven and earth: Modernism and reindeer in Chukota and Alaska. In *Northscapes: History, technology, and the making of Northern environments,* eds. Dolly Jorgensen and Sverker Sörlin, 174–194. Vancouver/Toronto: University of British Columbia Press.

Diubaldo, Richard. 1978. *Stefansson and the Canadian Arctic.* Montreal: McGill-Queen's University Press.

Exner-Pirot, Heather. 2012. Human security in the Arctic: The foundation of regional cooperation. *Working Papers on Arctic Security* 1(July): 2–11.

Fienup-Riordan, Ann. 2003. *Freeze frame: Alaska Eskimos in the movies*. Seattle: University of Washington Press.

Fitzhugh, William. 2009. 'Of no ordinary importance': Reversing polarities in Smithsonian Arctic studies. In *Smithsonian at the poles: Contributions to international polar year science*, eds. Igor Krupnik, Michael A. Lang, and Scott E. Miller, 61–77. Washington, DC: Smithsonian Institution Scholarly Press.

Francis, Jennifer A., and Elias Hunter. 2006. New insight into the disappearing Arctic sea ice. *Eos, Transactions American Geophysical Union* 87(46): 509–511.

French, Alice. 1992. *My name is Masak*. Winnipeg/Manitoba: Peguis Publishers.

Goldenberg, Suzanne. 2013. NOAA Report says Arctic sea ice is disappearing at unprecedented pace. *The Guardian*. 6 August 2013. http://www.theguardian.com/world/2013/aug/06/noaa-report-arctic-ice-climate-change

Government of the Northwest Territories. 2014. Northwest territories cumulative impact monitoring program. http://www.enr.gov.nt.ca/programs/nwt-cimp

Gruber, Jacob W. 1970. Ethnographic salvage and the shaping of anthropology. *American Anthropologist* 72(6): 1289–1299.

Gupta, Akhil, and James Ferguson, eds. 1997. *Anthropological locations: Boundaries and grounds of a field science*. Los Angeles: University of California Press.

Harrison, Faye V. 1997. Anthropology as an agent of transformation: Introductory comments and queries. In *Decolonizing anthropology: Moving further toward an anthropology for liberation*, ed. Faye V. Harrison, 1–14. Arlington: American Anthropological Association.

Hart, Elisa with assistance of Inuvialuit Co-researchers. 2001. *Reindeer days remembered*. Inuvik: Inuvialuit Cultural Resource Center.

Hinsley, Curtis. 1981. *Savages and scientists: The Smithsonian Institution and the development of American anthropology, 1846–1910*. Washington, DC: Smithsonian Institution Press.

Hossain, Kamrul. 2013. How great can a 'greater say' be? Exploring the aspirations of Arctic indgineous peoples for a stronger engagement in decision-making. *The Polar Journal* 3(2): 316–332.

Huntington, Henry P. 2013. A question of scale: Local versus Pan-Arctic impacts from sea-ice change. In *Media and the politics of Arctic climate change: When the ice breaks*, eds. Miyase Christensen, Annika E. Nilsson, and Nina Wormbs, 114–127. New York: Palgrave Macmillan.

Interdepartmental Reindeer Committee. 1933. 18 January 1933, G79-069, File 1-3, NWT Archives, Yellowknife, Canada.

Inuit Tapirit Kanatami. 2013. Press release: Canadian inuit leaders reject environmentalist campaign pitting indigenous peoples against Arctic resource development, 14 May 2013. https://www.itk.ca/media/media-release/press-release-canadian-inuit-leaders-reject-environmentalist-campaign-pitting

Inuvialuit Cultural Resource Center. 2016. Inuvialuit Pitqusiit Inuuniarutait (Inuvialuit Living History). http://www.inuvialuitlivinghistory.ca

Jacoby, Karl. 2001. *Crimes against nature: squatters, poachers, thieves, and the hidden history of American conservation.* Berkeley: University of California Press.

Jenness, Diamond. 1917. The copper eskimos. *Geographical Review* 4(2): 81–91.

———. 1921. The cultural transformation of the copper Eskimo. *Geographical Review* 11(4): 541–550.

———. 1923. A lecture delivered at the arts and letters club, by Diamond Jenness, Victoria Memorial Museum, Jan 9: 'Our eskimo problem'. Rudolph Martin Anderson Fonds, MG 30 40, vol 14, file 1-eskimos, Library and Archives Canada.

———. 1928. *The people of the twilight.* New York: The MacMillan Company.

Jenness, Stuart E. 1991. *Arctic Odyssey: The diary of Diamond Jenness, ethnologist with the Canadian Arctic expedition in Northern Alaska and Canada, 1913–1916.* Hull/Quebec: Canadian Museum of Civilization.

Jenness, Stuart E. 2011. *Stefansson, Dr. Anderson, and the Canadian Arctic expedition, 1913–1918.* Gatineau: Canadian Museum of Civilization.

Kevles, Daniel J. 1985. *In the name of eugenics: Genetics and the uses of human heredity.* New York: Knopf.

Kulchyski, Peter. 1993. Anthropology in the service of the state: Diamond Jenness and Canadian Indian policy. *Journal of Canadian Studies* 28(2): 21–50.

Levere, Trevor H. 1993. *Science and the Canadian Arctic: A century of exploration, 1818–1918.* New York: University of Cambridge Press.

Lovett, Laura. 2007. Men as trees walking: Theodore Roosevelt and the conservation of the race. In *Conceiving the future: Pronatalism, reproduction and the family in the United States, 1890–1938,* 109–130. Chapel Hill: University of North Carolina Press.

Lyons, Nastasha. 2007. *Quliaq tohongniaq tuunga* (Making Histories): Towards a critical inuvialuit archaeology in the Canadian Western Arctic. Ph.D. Deiss. Calgary: University of Calgary.

———. 2009. Inuvialuit rising: The evolution of inuvialuit identity in the modern era. *Alaska Journal of Anthropology* 7(2): 63–79.

Lyons, Natasha, Kate Hennessy, Charles Arnold, and Mervin Joe. 2011. The inuvialuit Smithsonian project: Winter 2009-Spring 2011. Report produced for the Smithsonian Institution, vol 1.

MacNeil, Jason. 2014. Inuit singer Tanya Tagaq's 'sealfie' photo supporting seal hunt sparks backlash. *Huffington Post,* 22 April 2014. http://www.huffington-post.ca/2014/04/02/inuit-tanya-tagaq-sealfie_n_5077203.html

McCannon, John. 2012. *A history of the Arctic: Nature, exploration, and exploitation.* London: Reaktion Books.

National Snow and Ice Data Center. 2012. Arctic sea ice extent settles at record seasonal minimum, 19 September 2012. http://nsidc.org/arcticseaice-news/2012/09/arctic-sea-ice-extent-settles-at-record-seasonal-minimum/

Nyhart, Lynn. 2010. Emigrants and pioneers: Moritz Wagner's 'Law of migration' in context. In *Knowing global environments: New historical perspectives in the field sciences*, ed. Jeremy Vetter, 39–58. New Brunswick: Rutgers University Press.

Okpik, Abraham. 2005. In *We call it survival: The life story of Abraham Okpik*, ed. Louis McComber, Life stories of Northern leaders, vol 1. Iqaluit: Language and Culture Program of Nunavut Arctic College,.

Palsson, Gisli. 2001. *Writing on ice: The ethnographic notebooks of Vilhjalmur Stefansson*. Hanover: University Press of New England.

———. 2003. *Traveling passions: The hidden life of Vilhjalmur Stefanson*. Hanover: Dartmouth College Press.

———. 2008. Hot bodies in cold zones: Arctic exploration. *Scholar and Feminist Online* 7(1).

Paul, Diane. 1995. *Controlling human heredity, 1865 to the present*. Atlantic Highlands: Humanities Press.

Peyton, Jonathan, and Robert L.A. Hancock. 2008. Anthropology, state formation, and hegemonic representations of indigenous peoples in Canada, 1910–1939. *Native Studies Review* 7(1): 45–69.

Piper, Liza, and John Sandlos. 2007. A broken frontier: Ecological imperialism in the Canadian North. *Environmental History* 12: 759–795.

Rainger, Ronald. 2004. *An agenda for antiquity: Henry Fairfield Osborn and Vertebrate Paleontology at the American museum of natural history, 1890–1935*. Tuscaloosa: University of Alabama Press.

Revkin, A.C. 2007. Arctic melt unnerves the experts. *The New York Times*, 2 October 2007.

Ryall, Anka, Johan Schimanski, and Henning Howlid Waep. 2008. Arctic discourses: An introduction. In *Arctic discourses*, eds. Anka Ryall, Johan Schimanski, and Henning Howlid Waep, ix–xxii. Newcastle: Cambridge Scholars Publishing.

Rynor, Becky. 2011. Indigenous voices 'marginalized' at Arctic Council: Inuit Leaders. Ipolitics. http://www.ipolitics.ca/2011/11/07/indigenous-voices-marginalized-at-arctic-council-inuit-leaders/

Sandlos, John. 2003. Landscaping desire: Poetics, politics in the early biological surveys of the Canadian North. *Space and Culture* 6: 396–409.

———. 2006. Where the scientists roam: Ecology, management, and Bison in Northern Canada. In *Canadian environmental history: Essential readings*, ed. David Freeland Duke, 333–360. Toronto: Canadian Scholars' Press.

———. 2007. *Hunters at the margin: Native people and wildlife conservation in the Northwest territories*. Vancouver: University of British Columbia Press.

Searles, Edmund (Ned). 2006. Anthropology in an era of inuit empowerment. In *Critical inuit studies in an era of globalization*, eds. Pam Stern and Lisa Stevenson, 89–101. Lincoln: University of Nebraska Press.

Simon, Mary. 2008. Sovereignty from the North. *Scholar and Feminist Online* 7(1): 1–3.

Smith, Neil. 2003. *American empire: Roosevelt's geographer and the prelude to globalization.* Berkeley: University of California Press.

Smith, Linda Tuhiwai. 2012. *Decolonizing methodologies: Research and indigenous peoples.* London: Zed Books.

Stefansson, Vilhjalmur. 1909. Northern Alaska in winter. *The Bulletin of the American Geographical Society of New York* 41(1): 601–610.

———. 1912. The Eskimo and civilization: Disease and death for the new Eskimo tribes with pauperization of those that chance to survive can be prevented only by a Quarantine which will allow the conditions of civilization but slow entrance to their territory. *American Museum Journal* 12: 195–203.

———. 1913. *My life with the Eskimo.* New York: MacMillan Company.

Stocking, George W. Jr. 1968. *Race, culture, and evolution: Essays in the history of anthropology.* Chicago: The University of Chicago Press.

———. 1987. *Victorian anthropology.* New York: Collier Macmillan.

———. 1992a. The boas plan for the study of American Indian Languages. In *The Ethnographer's magic and other essays in the history of anthropology*, ed. George Stocking, 60–91. Madison: University of Wisconsin Press.

———. 1992b. Ideas and institutions in American anthropology: Thoughts toward a history of the interwar years. In *The ethnographer's magic and other essays in the history of anthropology*, ed. George Stocking, 114–177. Madison: University of Wisconsin Press.

Stuhl, Andrew. 2013. The politics of the 'new North': Putting history and geography at stake in Arctic futures. *The Polar Journal* 3(1): 94–119.

———. in press. The experimental state of nature: Science and the Canadian reindeer project in the interwar North. In *Ice blink: Navigating Northern environmental history*, eds. Brad Martin and Stephen Bocking. Calgary: University of Calgary Press.

Swayze, Nancy. 1960. *Canadian Portraits: Jenness, Barbeau, Wintemberg: The man hunters.* Toronto: Clarke, Irwin, and Company Limited.

Warren, Louis, ed. 2003. *American environmental history.* Oxford: Blackwell Publishing.

Wheeler, Valerie. 1986. Traveler's tales: observation on the travel book and ethnography. *Anthropological Quarterly* 59(2): 52–63.

Wormbs, Nina. 2013. Eyes on the ice: Satellite remote sensing and the narratives of visualized data. In *Media and the politics of Arctic climate change: When the ice breaks*, eds. Miyase Christensen, Annika E. Nilsson, and Nina Wormbs, 52–69. New York: Palgrave Macmillan.

Worster, Donald. 1994. *Nature's economy: A history of ecological ideas*, 2nd edn. New York: Cambridge University Press.

CHAPTER 3

Petro-images of the Arctic and Statoil's Visual Imaginary

Synnøve Marie Vik

The Norwegian oil company Statoil is heavily invested in the extraction of oil and gas in the global North, and is equally committed to further exploring and developing fossil fuel resources in the Arctic. The company will depend on Arctic resources to secure a supply to a growing global energy demand. So far, they operate the producing gas field of Snøhvit (Snow White) in the Barents Sea and the northernmost operating liquid gas facility in northern Norway, as well as being a partner in several fields off the coast of Canada and Alaska. The company also explores fields in Greenland. In addition, they have major exploration deals in the Russian part of the Barents Sea and in the Okhotsk Sea, and licenses in Arctic waters in Norway, Russia, Canada, the USA and Greenland.

Naturally, this planned production carries ecological consequences, both in terms of the hazardous long-term environmental effects of continued reliance on a global economy based on fossil fuels presented in the Intergovernmental Panel on Climate Change, but also in connection with

S.M. Vik (✉)
Department of Information Science and Media Studies, University of Bergen, Bergen, Norway

© The Author(s) 2017
L.-A. Körber et al. (eds.), *Arctic Environmental Modernities*,
DOI 10.1007/978-3-319-39116-8_3

short term, local environmental effects due to possible oil spills. Statoil assures that its production is at the forefront of technology and completely safe, based on research, development and their long experience in challenging climates. The government of Norway owns 67 % of Statoil, and revenue from the company has been an important financial backer for the Government Pension Fund of Norway—previously the Petroleum Fund of Norway—the largest pension fund in the world. Yet, at the same time Norway frames itself as an environmental frontrunner. And the Norwegian Constitution §112 states that all citizens are entitled to a healthy, sustainable environment, that resources are to be distributed in favor of future generations and that citizens are entitled to knowledge on the effects of planned and ongoing exploitation of nature (Constitution of Norway 1992/2014).

Norway and Statoil's conflicting interests with regard to wealth and ecological safety make up the dispute over which the company's visuality is created. This chapter engages the theory and history of visual culture to analyze the ways in which images of Statoil's operations fuel dominant petro-narratives to gain dominance in the public imagination concerning the conflict between economic and environmental interests. It is certainly no surprise that the environmental impact would be minimalized in an oil company's PR material. An in-depth analysis of the imagery can however contribute to an understanding of the processes of trivialization of heavy industry in the Arctic.

When considering the Arctic future we must come to terms with its status as a highly covetable chamber of natural resources that sooner or later will be extracted, with great ecological consequence for both the local and global environment. While the mechanisms behind the decision making affecting the future of the Arctic are a complex network of powerful ideological, economic and not least the geopolitical agendas, we need to recognize the role of the Arctic imaginaries in fuelling these agendas. In our increasingly visual world of fleeting digital images, images exert a strong influence over our petro-narratives, bestowing great power to those who control them. Like any operative in the global market and political arena that is the oil business, Statoil is highly aware of the potentially powerful relations created by its visual identity, and widely distributes its own photographs of oil platforms, oil sand and fracking sites, and exploration sites in the Arctic. Their External Image Archive (Statoil ASA 2014) is but one example. These images, together with technical drafts and drawings, and generic business and industry photographs, constitute the company's visuality, a visuality that plays a critical part in the petro-narratives of the future of the Far North. This visuality confronts and influences the way we perceive the Arctic in terms of landscape, its aesthetic and its environmental challenges.

Statoil's imagining of the Arctic may be understood as a visual colonization of the Far North, contributing to a shift in the popular understanding of the Arctic and its resources. Seen (and romanticized) as a desolate, untouched and sublime landscape, the realm of indigenous peoples and polar explorers such as Fridtjof Nansen and Roald Amundsen, the Arctic now becomes the "promised land" of natural resources—oil and gas—that must be dug out. This extraction will alter the landscape, either directly through drilling and spills, or indirectly through rising temperatures. In this chapter I argue that the idea of the accessibility of the resources is made possible and believable by the ongoing visual imperialism of the oil companies, in this case Statoil. I use the term "imperialism" in a broad sense as the practice of extending power and authority over a territory.

The images of Statoil's operations, the landscapes they are situated in, and the relationship between the industrial operations and the nature in which they are embedded, contribute to the construction of the company's visuality. An awareness and understanding of the visuality of the oil business is crucial at a time when the resources of the Arctic are only starting to be explored. Statoil is involved in a creative, visual place-making where the man-made structures and technological edifices establish a visual place in an otherwise anonymous landscape, framing their operating sites as identity-less spaces, where nature is *backgrounded*. The term "backgrounding" originates in gestalt theory, but in this sense derives from eco-feminist usage of the term that draws out the power dynamics involved in our relationship to nature. Philosopher Val Plumwood described how, in Western culture, nature, like women, are backgrounded when modern capitalism exercises a denial of dependency on nature:

> One of the most common forms of denial of women and nature is what I will term backgrounding, their treatment as providing the background to a dominant, foreground sphere of recognized achievement or causation. ... What is involved in the backgrounding of nature is the denial of dependence of biospheric processes, and a view of humans as apart, outside of nature, which is treated as a limitless provider without needs of its own. (Plumwood 1993, 21)

The critique Plumwood formulates may be seen in relation to Nicholas Mirzoeff's notion of countervisuality, developed in his analysis of visuality, authority and power (2011). Mirzoeff's theory of a counterhistory of visuality in colonial history points out how power creates a *standard operating procedure*: an aesthetic that appeals to us, that seems "right," and thus encourages us to move on, since "there is nothing to see here." This aesthetic is what he considers to be the regime that controls our visual culture. By claiming "a right to look" and look again, we can conceptualize these forms of visuality and establish a *countervisuality* through the images presented by power, gauging the political implications of their aesthetic.

Mirzoeff's usage of the term "visuality" differs from the most common usage of it, where it is seen as a collection of images and artifacts, or a realm of experience. To Mirzoeff visuality consists of certain visual configurations that legitimize institutions of power and naturalize their cultural authority. He refers here to visuality as the semiotic constructions of imperialism as formulated by the nineteenth-century historian Thomas Carlyle. While realizing the significant differences between the pervasive institutions of the imperialism of bygone eras and the smaller contemporary institutions of capitalism—such as Statoil—I nevertheless hold that it is worthwhile employing visuality in this sense in my attempt to conceptualize Statoil's visuality in the North.

Following Mirzoeff, we see how the images of Statoil's oil and gas facilities are designed to give the impression of a standard operating procedure. Claiming our right to see the realities of the exploitation of non-renewable energy resources, a democratic politics emerges, in the sense that we make environmental issues part of the discussion on future energy sources. The right to look contests the right to exploit nature and jeopardizes the future of this planet. Claiming the right to look opposes autocratic authority and puts a countervisuality into play (Mirzoeff 2011, 29).

Statoil's press photographs of oil sand sites in Canada and fracking sites in the northern USA reach a wide audience through diverse media: news, industry press and PR. These sites are not included in the geographic definition of the Arctic, but they are representative of Statoil's visuality, especially for new ventures, which may tell us a lot about its approach to depictions of oil extraction and the surroundings. Statoil has been heavily criticized for their activities by environmental NGOs, perhaps most heavily for their oil sand project in Canada, a project that is detrimental to the surrounding landscape and has become a hot topic in Norwegian politics. The company's images of these sites, however, tell us nothing about this destruction. Fracking for shale gas is heavily disputed throughout America, but to no great surprise Statoil's images of their sites reveal no conflict.

In the following, we will see how the dominant aesthetic of Statoil's production sites globally creates a visuality that serves as an important backdrop to its endeavors in the Arctic. We should bear this visuality in mind when viewing their ongoing and planned operations in the Far North. The company's visual petro-narrative fuels our mental image of the future of the Arctic.

STATOIL'S VISUALITY

Statoil has had several images taken of their oil sand sites in Canada. One typical example is an air photo of the Leismer site in Alberta, Canada, taken by the Norwegian photographer Helge Hansen (2011). The site is photographed at a diagonal angle from a bird's-eye view. We see a square, flat lot with a few buildings and other structures on it, some identifiable as barracks, silos and hangars. A dirt road leads to the site. The ground is mostly mud, here and there some grass remains. The colors are muted and neutral. A small artificial-looking pond sits in the middle of the site. A closer look reveals a few cars and trucks. Large trees are clearly visible at the outskirts, and the surrounding forest seems to stretch into the distance, untouched. In the background we can make out a small lake. The whole scene seems like your average industrial site. It is difficult to distinguish exactly what kind of work goes on here. Overall the site seems well organized but eerily quiet. We cannot see any ongoing activity, trucks moving or workers walking around. Importantly, the site seems rather small, when in fact it covers a great deal of land (Fig. 3.1).

The landscape is backgrounded in this image, in the banal sense that the plant is in the foreground, but also in the sense that the plant is sharply distinguished, where the landscape is seemingly left untouched except for the confined and orderly area covered by the plant. This is even more evident in images of a shale gas production site in Pennsylvania (Hansen 2010). The site is portrayed as situated in the middle of a vast area of green forest, itself covering only a small section of the landscape. Hansen is in a sense performing a mapping of an area that the company is exploring, but it is as if the mapping is intended to keep the area in question under the radar. We are not supposed to get a feeling of where exactly this is. It is literally in the middle of nowhere, with no landmarks such as buildings or recognizable mountains in sight. The photographs fuel our mental image

Fig. 3.1   Statoil, Leismer, Canada. Photo: Helge Hansen, Statoil; courtesy Helge Hansen/Statoil

of these sites as being remote, unnoticeable and insignificant, and of the consequences of their activities as trivial (Fig. 3.2).

At the outset, Statoil's images of fracking seem to be of another kind, as they oftentimes portray the landscape with an eye for the beautiful. Surely these sites are supposed to be seen, the landscape even admired? The images showing hydraulic fracturing for shale gas in Williston, North Dakota, photographed by Ole Jørgen Bratland for instance, are taken during a stunning sunset over the serenely flat landscape of North Dakota (Bratland 2012a). This fracking site appears to be merely a trifle in the vast landscape, a technical structure that does not interfere with anything. Fracking occurs underground, and so the activity above ground does not tell us much of what is going on. Nor does the production equipment take up much space. A dirt road is seen in the foreground, dust swirling to the left. The colors are warm reds and browns. The site is located to the left in the image, the sun is setting to the right, creating a dual focus for the eye, leaving the central perspective in the distance: the site is not the center of the image, the landscape itself is. This perspective is repeated in several photographs of the same site taken by Bratland. The sunset turns

Fig. 3.2   Shale gas production in the hills of Pennsylvania, USA. Photo: Helge Hansen, Statoil; courtesy Helge Hansen/Statoil

our interest away from the activity and to the landscape. In other images the main focus is the dirt road continuing into the distance, diverting attention from the site altogether, focusing instead on the vastness of the landscape (Bratland 2012b). In yet another photograph the focal point is divided between the fracking tower and a hill, almost pushing the technical construction out of the picture and out of the landscape (Bratland 2012c).

In these images, contrary to the oil sand sites, nature is seemingly foregrounded. In an information video of their fracking operations, there are even grazing cows in the front of the site. The images come off as landscape photographs with a fracking structure casually part of the motif. But just as in the oil sand images, the fracking images involve an aesthetic that is designed to be part of the standard operating procedure, telling us to keep going, as there is nothing specific to see here. As visual culture scholar W. J. T. Mitchell points out when writing on the power of landscapes, a landscape is commonly not defined in terms of its specificity, but is rather the overlooked, not the looked at. "Look at the view," we typically say (Mitchell 2002, vii). In a similar fashion, Statoil invites us exactly to look at the view, and not the details of the landscape.

As landscape pictures these images resemble the perspective seen in photographs of the great American wilderness—the promised land—taken by Timothy H. O'Sullivan between 1860 and 1880, images that, as Joel Snyder puts it:

> provide visual, photographic proof of the unknown character of the land and imply the need to gain power over it by coming to know it. The job is to probe the territory, subject it to scientific examination, thus understanding what it can tell us about its past and how it can be used in the future. (Snyder 2002, 199–200)

Statoil's operations engage in such probings, and the images, showing the company's uneventful mastery over the landscape, are arguments to let them carry on that effort on our behalf.

Statoil's self-imaging as master of the landscape is of course grounded in their long-time position as operator of oil rigs in the North Sea. The traditional, and in a Norwegian context iconic, photographs of their oil rigs at sea adhere to the same aesthetic as the fracking and tar sand sites, relying on our trusting them to proceed with standard operation procedure. The oil rigs are nearly always the focal point of the image, the sea portrayed as a non-eventful background. More often than not they are photographed in spectacular weather conditions, often gleaming in the warm light of the setting sun (Hagen 2011). The images of the cargo ships transporting the oil and gas globally adhere to the same aesthetic, where the everyday routine of the shipping business is made evident through enhancing the formal aspects of the cargo ships, and situating them in calm sea—again, often glowing in the sun (Nesvåg 2011).

Since production in the Arctic is still at a planning stage, visualizations of their future extraction are inevitably merely imagined projections. Images presented online and in strategic documents are therefore even more vague than usual. The visual material presented in "The Final Frontier: Statoil's Arctic Exploration Portfolio and Strategy" serves as an example (Hansen 2012). Strictly considering the photographic material used for illustrations, the document offers a curious perspective on their plans for the Arctic. The document contains three larger photographs and five smaller ones used as background for text. Of the three larger ones, one is of the processing facilities at Melkøya, where natural gas is liquefied to be exported globally, which is portrayed much in the same fashion as Statoil's on-land sites elsewhere, modestly situated in an otherwise beautifully captured landscape (Hansen 2012, 12). Contrary to the images of existing facilities, which for the most part are devoid of humans, the other two images portray Statoil employees in working gear staring into the sea. One photograph is featured twice, at the front and last page, and depicts two men walking away from the camera, towards the edge of what looks like a rig and towards the sea and the setting sun (Hansen 2012, 1, 17). The men are at the center of the picture, but the illuminated horizon draws the gaze of the viewer towards the open sea. As beholders, we join the anonymous rig workers on their open quest into the unknown. The third image breaks with the rest of Statoil's visuality. It features two men in the bottom right corner, standing on the deck of a boat, gazing into the icy ocean (Hansen 2012, 16). The picture is special in two regards, in that we can see the face of one of the men, and in that the picture has been given a subtitle: "Arctic research trip to East Greenland 2012" (Hansen 2012, 16). The proximity to the two men, the narrower focus and their individual appearance, together with the categorizing subtitle, give the picture a sense of actuality lacking in the others. When presented to potential shareholders, the expression of authority changes.

Move On, There Is Nothing to See Here

Statoil promotes a sense of normality in their images, diverting our attention with a documentary aesthetic, undermining the position of critical voices. According to Mirzoeff, visuality is neither the collection of images one normally thinks of, nor a realm of experience, but rather the "standard operating procedure" by which the visual operates. Mirzoeff relies heavily on the writings of the philosopher Jacques Rancière and his notion of the "police order" and its ruptures. Very simply put, Rancière distinguishes between the order of the police, the system we live under in the everyday, with its distribution of power and positions, and politics, which points to events that break with the police order, as in the situation where those who have no voice in public start speaking. Politics, which for Rancière is always a question of democracy, is in this sense a clear break with the idea that positions in society are fixed and that there exist predispositions for a given order of things (Rancière 2010, 30–31).

Mirzoeff articulates visuality and countervisuality in line with the police order and the political event, and his aim is to highlight the ways in which countervisuality may break the supremacy of the genealogy of visuality. He does this by identifying three "complexes of visualization," namely plantation slavery, imperialism and the modern military-industrial complex, linking them to power through techniques of classification, separation and aestheticization. In this way the visual is named, categorized and defined; it is separated and segregated into groups, preventing those who are visualized from becoming political subjects; and it is further rendered in a way that is pleasing, that is aesthetic. Visuality is according to Mirzoeff "that authority to tell us to move on, that exclusive claim to be able to look" (Mirzoeff 2011, 2). Mirzoeff bases his argument on Jacques Rancière's differing account of subjectivation from that of Louis Althusser. While Althusser's famous concept of "interpellation" is based on authority addressing the subject "Hey, you there!" Rancière lets authority usher the crowd away, "Move along! There's nothing to see here!" (Rancière 2010, 37).

This is the visualization of history, one that manifests the authority of the visualizer, and that has come to be seen as routine, customary. All this depends on a submissive class (us) which adheres to the work that needs to be done, without seeing. Mirzoeff challenges this by claiming "the right to look" as a countervisuality that "claims autonomy from this authority, refuses to be segregated, and spontaneously invents new forms" (Mirzoeff 2011, 4).

Seeing the right to look as a way of democratizing democracy, in that the looking interconnects with the right to *be seen*, might contribute to a useful analysis of Statoil's images. Statoil, claiming authority, controls visuality. This visuality is a process, or history in the making. Mirzoeff suggests that we should be treating visuality as a discursive practice that has material effects. We have a choice, a choice between moving on or claiming that there *is* in fact something to see (Mirzoeff 2011, 5). In our case, our gaze meets nature, and thus invites it into the dialogue. Nature does not have an agenda of its own, which can be seen to be in conflict with humanity's interests; rather it is that we are nature, and that the backgrounding of nature excludes the central precondition for our existence. This exclusion carries with it great implications for our discourse on the politics of energy recourses. If we are able—through the right to look—to claim "the right to the real," this might function "as the key to a democratic politics" (Mirzoeff 2011, 4). By studying the images of Statoil we are exercising a right to the realities of exploitation of non-renewable energy resources, which is key to a democratic politics in the sense that we make environmental issues part of the discussion on future energy sources.

The mode of visibility, the link between authority and power that is displayed in these images, bears resemblance to, and is in fact hereditary to, traditional landscape paintings, though it interestingly also mimics the layout of the sugar plantations. Where the plantations exploited human resources, oil companies exploit natural resources. As the plantation owners empowered themselves with this authority to exploit their slaves, so too it could be argued that Statoil empowers itself with the authority to exploit nature. The visuality of the plantation system versus the visuality of oil today is a parable in power and imperialism, facilitated by globalization. The plantation owners treated humans as nature, while nature was and is considered to be opposed to culture, and consequently to be ruled by human beings. The images of oil sand extraction and fracking mimic a long tradition of visualizations of "culture over nature," where culture equals a Western civilization that tends toward "order" and "perfection." The order of Western culture was seen to be *right* because it seemed perfect, its aesthetic qualities validated it ethically. Statoil's images make use of this connection between power and aesthetics.

Like the Haitian plantations that transformed the disagreeable landscape through European oversight, so too can we claim that Statoil "orders" and classifies nature from culture in their quest to oversee the oil landscapes. And in the same way that the plantation structures rarely changed from country to country, but instead were set up in the same way, using drawings and manuals across the colonies, so too does the aesthetic of oil images rarely change. An oil site is portrayed in the same way wherever Statoil ventures. Mirzoeff remarks on an illustration of a plantation, published in 1667 by the missionary Jean-Baptiste Du Tertre:

> Even the landscape attests to the transformation wrought by European oversight on the indigenous condition of the land, which Du Tertre called "a confused mass without agreement." The mountains and indigenous wilderness in the background of his image give way to the regularly divided and organized space of plantation. (Mirzoeff 2011, 52)

Similarly, Statoil performs an oversight of capital as a regime of power, where the wilderness is presented in the background, while the oil sites are presented as regularly divided and overall very organized places. Importantly, this is a form of mapping of the landscape that also mimics that of the plantations: mapping "rendered a colonial space into a single geometric plane" and "developed the distinction between 'cultivated' and 'empty' space that motivated settlement into a determining principle that organized and aestheticized perception" (Mirzoeff 2011, 58). This relationship between cultivated and empty space is still very much relevant as Statoil moves into the Arctic, even if it is changing.

Imperialism of the Global North

To understand the relationship between Statoil's oil extraction and the images of their operating sites, we might also engage with Mitchell's description of the concepts space, place and landscape as "a dialectical triad, a conceptual structure that may be activated from several different angles. If a place is a specific location, a space is a 'practiced place,' a site activated by movements, actions, narratives and signs, and a landscape is that site encountered as image or 'sight'" (Mitchell 2002, x). The framework for any discussion about oil and nature today is the free movement of capital in the global market economy, a process that neglects and obliterates any difference between the involved elements. Statoil's images situate the operating facilities in non-specific spaces. Following Mitchell's dialectical triad, we recognize how Statoil's sites are not places in terms of place as a specific location, but rather the practiced place of an indefinite space, or in other words a non-local locality. The production is inevitably local somewhere, even if this somewhere remains afar and unapproachable by us, and so the consequences of production are local too. The generic visuality of Statoil's images weakens the impression of locality conveyed.

When it comes to phenomena such as global warming however, "locality is an abstraction," as Timothy Morton argues (2013, 47). There is no such thing as the local, only non-locality, that is, local events are always connected to events and processes that are going on outside of any understanding of the local. The economic network that Statoil is a part of is global, and so the fossil fuel that the company produces is a global commodity. The atmospheric changes that the fossil fuels continue to contribute are by definition global phenomena. Statoil's images navigate these waters carefully, by acknowledging *a* locality, constructing a place for oil extraction to happen, while at the same time not really granting these places an existence. This seems to be the perfect visualization of a practice that must downplay its environmental impact locally and globally, while emphasizing the reality of its product.

A visual parallel can be found in military sites, as seen in the work of American artist and geographer Trevor Paglen. Photographing the secret sites of the American military in the Nevada and New Mexico deserts (Paglen 2012), and NASA's sites in West Virginia (Paglen 2010), Paglen reveals them to be the true landscapes of globalization: classified landscapes for covert operations. Paglen's photographs depict control towers, surveillance sites, hangars, and vehicles, shot with telephoto lenses. Rebekka Solnit argues—congruently with Mirzoeff—in an essay on Paglen's photographs, that they rupture the invisibility of a society at war with itself (Solnit 2010, 9). "Invisibility," she points out, "is in military terms a shield, and to breach secrecy is to make vulnerable as well as visible" (Solnit 2010, 10). The sites of military and space programs are highly secretive and not supposed to be accessed or seen by anyone without clearance—that is, the proper authority to see. The government wants these sites to be invisible and out of reach. Paglen chooses to see, asserting his right to look (Fig. 3.3).

While not implying that Statoil is as secretive as the American military or NASA, it is nevertheless striking that they have a shared visuality, and

Fig. 3.3   Trevor Paglen, *They watch the moon* (2010). Courtesy the artist

that Statoil in a number of instances frames its operations in the same way as Paglen claims the secret sites in the American wilderness do. Paglen's photographs are best seen as examples of a dissensual gaze, or the countervisuality to the visuality of a militarized situation, insisting on the visibility of that which is not to be seen, but forced into a distant perspective. Statoil's images are, on the other hand, a willed distancing, letting their sites be seen, but only as something that should not be bothered with.

When entering new territory in the Arctic, Statoil's visuality changes, a change pointing towards a crisis in the very visuality. Given that the main objective of visuality is to pass itself off as the natural state of things, a visuality that is noticed points towards the crisis of that visuality (Mirzoeff 2011, 6). Of course, the ones who notice are those who are especially trained to read images, or have other means of placing the images in a context that imbues them with a different meaning than what was intended. Detecting visuality "requires educated eyes" (Solnit 2010, 15). But this particular change is in a sense contributing to the education of the beholders: Statoil's move into the Arctic is made possible by the very global warming to which fossil fuels contribute, while the still harsh conditions of the Arctic demand very different installations from the ones Statoil have usually relied on. These installations are "invisible."

The Arctic is for the most part still not developed when it comes to oil and gas extraction; however, one exception is Statoil's first sub-sea installation, the Snøhvit field at the bottom of the Barents Sea, currently extracting gas from nine wells, planned to be 20. The sub-sea installation is invisible at the surface of the sea landscape; the gas is then transported by a network of tubes spanning 143 kilometers onto highly visible processing facilities on land on Melkøya. Statoil's future production facilities will for the most part be similar sub-sea installations, which present new questions related to their visuality.

Since the company's ongoing and future venture into the Arctic relies mainly on such installations, they cannot be photographed in the same way as their sites on land and standard offshore activities. This technological change strips them of the possibility of framing their extractive operations as standard operating procedure within and opposed to a larger wilderness. In this sense, a countervisuality arises, not only from our demand to look, and look again, but from the impossibility of backgrounding nature while taking pictures under water. This change may in fact help us see the metaphoric waters that we swim in a society built on the unsustainable extraction and use of fossil fuels. Statoil's crisis of visuality is a crisis of ecology.

IMAGES

Bratland, Ole Jørgen. 2012a. Williston, North Dakota. Statoil Image Archive. Archive number 0080117.

———. 2012b. Williston, North Dakota. Statoil Image Archive. Archive number 0080095.

———. 2012c. Williston, North Dakota. Statoil Image Archive. Archive number 0080110.

Hagen, Øyvind. 2011. Troll C platform. Statoil Image Archive. Archive number 0069223.

Hansen, Helge. 2010. Shale gas production in the hills of Pennsylvania. Statoil Image Archive. Archive number 0065997.

———. 2011. Leismer. Statoil Image Archive. Archive number 0072428.

Nesvåg, Øyvind. 2011. Njord field Njord B. Statoil Image Archive. Archive number 0069802.

Paglen, Trevor. 2010. They watch the Moon. Paglen.com. http://www.paglen.com/?l=work&s=theywatchthemoon. Accessed 20 Sept 2016.

———. 2012. Limit telephotography. Paglen.com. http://www.paglen.com/?l=work&s=limit. Accessed 20 Sept 2016.

Statoil ASA. 2014. "Statoil External Image Archive." http://fotoweb.statoil.com/fotoweb/statoil_Startpage.fwx.

WORK CITED

Hansen, Runi M. 2012. The final frontier: Statoil's Arctic exploration portfolio and strategy. *Science & Justice: Journal of the Forensic Science Society*. Stavanger: Statoil ASA. www.statoil.com/no/InvestorCentre/Presentations/2012/Downloads/Statoils%20Arctic%20exploration%20portfolio%20and%20strategy%2013%20Dec%202012.pdf. Accessed 20 Sept 2016.

Mirzoeff, Nicholas. 2011. *The right to look: A counterhistory of visuality*. Durham: Duke University Press.

Mitchell, W.J.T. 2002. Preface to the second edition of landscape and power. Space, place, and landscape. In *Landscape and power*, 2nd edn, ed. W.J.T. Mitchell, vii–xii. Chicago: The University of Chicago Press.

Morton, Timothy. 2013. *Hyperobjects. Philosophy and ecology after the end of the world*. Minneapolis: The University of Minnesota Press.

Norway. 1992/2014. *Constitution of Norway*. Available at: https://lovdata.no/dokument/NL/lov/1814-05-17/KAPITTEL_6#KAPITTEL_6. Accessed 20 Sept 2016.

Plumwood, Val. 1993. *Feminism and the mastery of nature*. New York/London: Routledge.

Rancière, Jacques. 2010. *Dissensus: On politics and aesthetics*. London: Continuum.

Snyder, Joel. 2002. Territorial photography. In *Landscape and power*, 2nd edn, ed. W.J.T. Mitchell, 175–201. Chicago: University of Chicago Press.

Solnit, Rebekka. 2010. The visibility wars. In *Invisible: Covert operations and classified landscapes: Trevor Paglen*. New York: Aperture Foundation.

CHAPTER 4

Arctic Urbanization: Modernity Without Cities

Torill Nyseth

The urban Arctic in Scandinavia reveals a number of forms and particularities that are distinct to the Arctic and yet bear resemblances to urbanism in other parts of the world. Arctic cities can be seen as an urban paradox, challenging what we know and think about what urbanity means. Taking account of these seemingly simple observations, the objective of this chapter is twofold: it provides an overview of recent developments and features of Arctic urbanism with a focus on northern Scandinavia, and it raises questions that point towards alternative or more inclusive discourses of urbanity.

Despite the fact that the Arctic is often characterized as a place uninhabited by humans, urban settlements there are growing rapidly, demanding a change in our perception of the region as only inhabited by seals and polar bears. Processes of urbanization are extreme in some areas, linked to the hyper-industrialization that follows the exploitation of natural resources. This urbanization is driven by the need to develop the necessary soft infrastructure and "light institutions" to facilitate resource extraction, including knowledge centers, social and public services, research institutions, financial infrastructure, and a variety of amenities to increase the attractive-

T. Nyseth (✉)
University of the Arctic Institute for Sociology, Postboks 6050, 9037 Tromsø, Lanees, Norway

© The Author(s) 2017

L.-A. Körber et al. (eds.), *Arctic Environmental Modernities*,
DOI 10.1007/978-3-319-39116-8_4

ness of the place for its new inhabitants and investors. In the Scandinavian north, this infrastructure is being developed in regional capital cities such as Tromsø (Norway), Rovaniemi (Finland), and Umeå (Sweden), which function as case studies for this chapter. Smaller cities, like Alta and Hammerfest in Norway, and Kiruna in Sweden, are also growing, driven in part by the resource extraction industry. The role and function of these cities in their respective regions is also changing. Some of them seem to be "winning" the competition between cities, while others are stagnating. Some of them will be part of the transnational service network and its mobile middle class—a global network of finance, production, and economy—and have to adopt their infrastructures and individual characteristics to these forces and their modes of life.

These forces are transforming both the living conditions in the region and the cities themselves, what they represent, what goes on there, as well as the design of the cities. The Arctic city is being produced in an era of "Arctic-fication" and within the context of industrialization, cultural shifts, and the new politics of the High North. By "Arctic-fication" I mean the process through which current symbolic discourse about the Arctic is challenging old myths about the region and constructing new ones (Guneriussen 2012). The Arctic wilderness has been turned into something beautiful and spectacular, with no remaining trace of the older images of an unfriendly, dark, and inhospitable landscape. Arctic "magic" has recently been "discovered" and is now found in the most unlikely places and events. The Arctic has become a new frontier, a magical region, representing a prosperous future for industrial development (Guneriussen 2008). These cities are also multi-ethnic, diverse, and characterized by various forms of hybridity.

David Bell and Mark Jayne argue that small cities have been ignored by urban theorists for too long (2009). This is also the case with Arctic cities, which have not been on the research agenda until recently, even though processes of globalization and urbanization are profoundly apparent in the North (Dybbroe et al. 2010). Arctic cities are mostly small in scale and population, but represent different forms of urbanity, an urban pluralism. In the following, I look more closely into the particularities of Arctic urbanism, its different drivers, and some of the forms of its expression. What is the essence of the Arctic city? How ordinary is it, and how particular? How could it be described? What elements of urban life become important? Do we see other forms of urbanity emerging as a consequence of city growth, of the new industrial paradigm related to extractive industries, or because of the geographical specificities of the Arctic?

Notes About Scale and Roles

In the Arctic and sub-Arctic regions we find a hierarchy of cities within the urban structure that includes larger cities, smaller towns, and settlements. One may question the characterization of many of these places as cities. While there are some larger cities in the region—particularly like Murmansk in Russia, with more than 300,000 people—in the Scandinavian North, using classical definitions of the city as a dense, large, and socially heterogeneous, multi-functional, and mixed-use space with a specialized labor force (Wirth 1938/2003), there are mainly smaller settlements. In the sub-Arctic region we find medium-sized cities (50,000–100,000), such as Tromsø, Umeå, Oulo, and Bodø, and smaller towns (10,000) such as Hammerfest, Kirkenes, Narvik, Harstad, Alta, Rovaniemi, and Kiruna. In Iceland there is only one city, Reykjavik, which has the dominant position as the capital. In Greenland, Nuuk is defined as a metropolis, even though it only has 17,000 inhabitants. Even though many of these towns do not qualify as cities because of their size, some of them still "act" like major cities, particularly Nuuk. Its status as a metropolis is related to its role as the capital, and subsequently its role in the global economy.

In the Arctic region, urbanization needs to be reflected upon with regards to what in other regions would be considered to be rural settlements and small towns in size and functionality. Considering the forces reshaping these northern areas, it is important to understand that the changes we are witnessing signal the emergence of a new type of small-scale urban development. Despite the rapid urbanization of the Arctic, the Arctic city does not figure in mainstream theories of urbanization. Their functions differ from location to location, from regional cities with no decisive power regarding the political, administrative, or economic issues affecting them, to capital cities that play an important role in the national economy. They are not global cities if that means being command centers in the world economy (Sassen 1991) or centers of production and consumption of the advanced services connected to global networks (Castells 1996). The cities in the Arctic North are expected to be motors of regional growth, however, particularly these days; and the diversity that comes along with changes in the economy significantly enhances the potential for innovation and experimentation. The relationship between Arctic cities and their transnational and territorial contexts are far from clear in terms of political power, strategic decision making, economic relations, and networking because there is a non-horizontal, south–north hierarchy of relationships. The political and economic links are tied to the national capitals in the south, not towards other cities in the Arctic.

There are, however, important exceptions to this pattern. Kirkenes, located at the Norwegian–Russian border, has developed close connections to the cities on the Russian side (Viken and Nyseth 2009) since the opening of the border following the dissolution of the Soviet Union. The creation of the Arctic Council in the 1990s and the establishment of its secretariat in Tromsø in 2013 have led to extensive connectivity across the region. Arctic cities could thus be characterized through their degrees of openness to external influences and their level of connectivity. The combination of openness and exteriority, along with their territoriality, shape their dynamic potential. In several Arctic cities, potentials and opportunities for new development have emerged as a result of the interactions and intersections this connectivity enables. However, these cities are still viewed as fixed entities in a classical sense—relatively stable, nested, geographical areas defined by their export-oriented production (fishing, mining, oil, natural gas), exploited by transnational politics, economics, and business. This perspective signals the presence of a latent colonial tendency in understanding Arctic areas and cities.

Arctic Cities and Arctic Urban Nature

So what is an Arctic city? Using standard references to urban theory, it is perhaps easier to describe what the Arctic city is not, rather than what it is. At a time when half of the world's population lives in cities, the city is everywhere and in everything. It is difficult to identify what is not urban. According to Amin and Thrift (2002), we can no longer agree on what counts as a city, but we still think of cities as distinct places. It could be helpful to think of the urban as more of a process than as a fixed thing. The concept of "metapolis" coined by Francois Ascher (2007) could be helpful in describing Arctic urbanism. Metapolis can be thought of as the third modern urban revolution, constituted by a regional urbanism achieved through enhanced regional mobility, new regional agglomerations for production, knowledge, and economy, and regional connectivity through transnational networks of transportation and communication. Some studies indicate the existence of a certain urban "ethos," particularly among young people in the North (Beck 2004), but how is this urbanism expressed, practiced, and performed in Arctic cities? Is it marked by a particular spatiality? How is the urban space produced in these geographically isolated locations with harsh climatic conditions, with long distances between cities? What are the spatial drivers and outcomes within this area of small settlements in large landscapes?

The settlement structure in the Arctic has undergone a major process of transition in recent decades, with people moving away from small villages in order to settle in larger towns. Most of the population growth experienced in the Arctic region occurs in urban centers. Arctic nature represents an "empty" space in some of these areas, devoid of permanent settlements. People live in cities or smaller villages, densely populated pieces of land below huge mountains, as along the coast of Finnmark. Here there are scattered harbor towns and between them there is "nothing," only a harsh coastline making it impossible to build a house or take a boat to sea. In the fjords, remnants of a traditional way of life can still be found; household economies based on combinations of fishing and farming still exist in sparsely populated areas, far away from towns or cities. This way of life has recently declined dramatically, with young people moving away, leaving behind an aging population (Bæck and Paulgaard 2012). One of the spatial particularities of Arctic urbanism is that most of the population growth occurs in urban centers with distinct borders. They are like small urban spots in the wilderness, with practically no suburban areas and sparsely populated surroundings. Because of their density, and economic and cultural diversity and vitality, these centers offer an "urban way of life" compared to their surroundings (Munkejord 2009). They function as centers of administration and other services, of knowledge, and of finance to an increasing degree.

Arctic cities are located close to nature and *are* nature to some extent. There is a close relationship between culture and nature in these cities. In post-modern industries like tourism that are springing up in these cities, it is this seemingly "empty" space of nature that attracts people and makes the development of a tourist industry possible (Viken 2011). For most of the tourists, the city is a place to rest in between the exploration of the wilderness. The city, with its urban forms of life, represents an exception from the wilderness, a stark contrast to other parts of the world where urban sprawl makes it difficult to distinguish one city from the next, and where the only "nature" is the city park. An urban "ethos" (Beck 2004) is followed by intimate relations to the non-urban through extreme performances in "the wild." People living in Arctic cities spend much of their free time in nature, all year round. This intimate relationship to nature may very well be one of the elements that make Arctic cities livable; a particular quality of these cities is the easy access to extreme sport activities such as skiing in rough mountain terrain or mountain climbing. These cities attract a growing segment of highly competent urban athletes seeking adventure, challenges, and risks.

Nature is also a part of the urban life of these cities. Sledding dogs are the "urban foxes" of the North. In Alta, Norway, the Finnmark Race has become a huge global sporting event, gathering dog-sled teams from all over the Arctic region during the week-long event (Granaas 2015). The city becomes a meeting place between dogs and people—between the human and the non-human. These relationships with nature challenge the spatial divisions between the civic and the wild, producing what Hinchliffe and Whatmore (2006, 124) call "heterogeneous urban inhabitants." While "urban foxes" in the UK symbolize the adaptation of nature to the city, the urban Arctic is a relational practice where human activities are urgently dependent on their adaptation to a nature that not only surrounds them, but also permeates the urban settlements. For example, in Hammerfest, the fences there are not only designed to keep reindeer out, but function at the same time to keep its urban inhabitants inside.

ARCTIC DIVERSITY: INDIGENOUS, MULTICULTURAL CITIES

Arctic cities are multicultural in a very specific sense. They are meeting places for several different ethnic groups, including local indigenous peoples and a diverse range of nationalities from all over the world. In Tromsø, for example, there are more than 130 different nationalities present. Tromsø also defines itself as an indigenous city, home to the largest population of Sámi in Norway outside the Sámi core districts. Kirkenes has become a bilingual city as a consequence of its large Russian population. Historically the population of the Norwegian north has been a mix of three "tribes"—the Sámi, the Kvens (Finnish people), and the Norwegians—and, in some parts, also Russians. These areas have been the home to people from different nationalities and cultures for several hundred years. Russians, Finns, and Sámi have shared the land in the eastern part of Finnmark since the seventeenth century. Russian immigration, on the rise since the fall of the Soviet Union, is therefore not a completely new event, but represents a historic continuity.

Many Arctic cities are located on indigenous land, even though this territory has not been recognized as indigenous by the national majorities of these countries. Decolonization is an ongoing process, creating a complex ethnic fabric in the city. The majority of indigenous people, whose identities and traditional cultures are linked to remote, rural places, are today living their lives in urban centers. This increasing urbanization of the indigenous population is a recent development, related to both the reclaiming of identity and to increased mobility from rural to urban areas. Therefore, multiculturalism has a long history in the region, but its current forms are new. This is a result of the increased mobility into and out of the region, including temporary mobility, like Russian fishermen visiting Kirkenes on a short-term basis, and the influx of travelers of all kinds including tourists, merchants, and temporary workers (Viken and Swencke Fors 2014).

If cities are key sites where new identities are formed (Sassen 2012), this is certainly happening in Arctic cities. Experimentations with indigenous identity is an ongoing process, with the emergence of new cultural practices that are non-existent in the rural communities from where they originate. Rejecting the historical imaginary of indigenous people being "out of place" in urban contexts, they are now claiming the cities as their own with the same legitimacy as anyone else, injecting them with new forms of hybrid identity (Nyseth and Pedersen 2014). A century ago, assimilation into the majority culture was the only option for indigenous migrants to the cities, which stripped them of their cultural distinctiveness as a consequence. Today's Arctic cities seem able to include indigenous cultures in the production of a multicultural city image, although not entirely free of conflicts and negotiations.

At present, Sámi institutions are "rooting" Sámi into everyday life in the cities. In a comparative context, there are few resemblances to studies from other parts of the world where indigenous groups are highly urbanized, as in Canada, the USA, and Australia. In these places, the focus is mainly on marginalization, poverty, drug abuse, and homelessness (Kishangani and Lie 2008). These issues are hardly relevant in the context of the Scandinavian welfare state. Because Arctic cities are heterogeneous when compared to the more uniform subcultures in the Sámi districts, these cities are particularly interesting to study as meeting places between cultures. There are differences and variations between Arctic cities in terms of how indigenous identities are expressed and struggled over. In Tromsø, for example, urban modernity seems to enhance the personal expression of Sámi identity, while in Rovaniemi, Sámi culture is highly commercialized by the tourism industry and put on display in very odd ways in order to attract more visitors. In parts of Finnish Lapland, the commercialization of Sámi symbols and traditions has been much more prevalent than in Norway or Sweden (Viken and Pettersson 2007). The Sámi revitalization processes have played out differently in each national and urban context because of the different policies surrounding the creation of Sámi institutions and policies regarding the Sámi language that together have played a role in shaping urbanity.

Urbanization and Hyper-Industrialization

The driving forces behind urbanization in the North differ somewhat from the global trend of the post-industrial city with its symbolic economy based on "soft" competences like the culture industry, and network and information technology. The urbanization of the Arctic is driven in part by hyper-industrialization (Benediktsson 2009). The "opening" of the region due to climatic, technological, economic, cultural, and political change has turned it into a "hot" area of geopolitical focus. This changed geopolitical environment makes the Arctic appear to be a prosperous region for advanced industrial production. Many cities were established as industrial ones, like the mining cities of Kirkenes, Kiruna, and Narvik, which came into existence around 1900. Others are facing a new phase of industrialization as a consequence of the global race for fish, oil, gas, and minerals of all kinds. Hammerfest is a typical example of a city that is being transformed from a small one, based on fish production, to one with a petroleum-based economy within global networks.

The economies of these cities are not only characterized by the export of natural resources at all levels of the production process, but also by the export of capital. Although these economies are highly globalized, little if any of the export income is circulated back into the local economy. There are almost no locally owned businesses involved in any of these industries. Take the coal mine in Kirkenes as an example. Until 1996, the mine was operated by a state-run company, and now it is owned by an Australian mining company. A global marketplace has brought in new actors from faraway countries, creating new relational networks between places, firms, and individuals.

Some of these new industrial towns look like large construction sites, where most of the population increase is due to a multinational army of construction workers. Others are more like industrial "working camps" based on "fly-in, fly-out" concepts of the workforce. The new residential houses that are being built are constructed rather cheaply and without much consideration to either design or site location. The large silos filled with cement that line the piers are visible signs of the ongoing construction. The Russian Arctic is developing at high speed, as are parts of the Swedish and Finnish North, and the international mining companies are knocking on the doors of the Norwegian Arctic as well. As mentioned above, in Kirkenes the mine has been reopened by international capital after it was closed down in 1996. It is located at the now less rigidly controlled Russian border, and there is a strong Russian presence there. Kirkenes is becoming the center of a highly intensified exchange of people, goods, and knowledge across the border between Norway and Russia (Viken and Nyseth 2009; Viken and Swencke Fors 2014). Despite this transnational exchange, mining still imprints itself on the town's economy and image.

The Cultural Economy, Politics, and Design of the Modern Arctic City

The cultural economy of the urban Arctic also has its specificities. Festivals like the Tromsø International Film Festival and the Barents Spectacle in Kirkenes, both of which take place in January during the darkest period of winter, attract a large number of visitors. Umeå was appointed the European City of Culture in 2014, and Tromsø bid for both the 2014 and 2018 Olympic Winter Games—without success, although these attempts display an ambition and a willingness to take on the responsibility of a huge sporting event. All of these events challenge old myths about the region as underdeveloped and provincial, and produce new symbolic meanings that provoke images of the exotic, magical, and spectacular (Guneriussen 2008). It is not clear how this cultural economy might be affected by the new and expanding industrial development. Culture might prove to be a contestable and conflicted aspect in the branding of these cities, where image is everything.

During the Cold War, security considerations defined national policy priorities for the North, influencing the economic bases of cities like Kirkenes on the Norwegian side and Murmansk on the Russian side of the border. State support for the mining company Syd-Varanger A/S in Kirkenes had primarily military instead of economic motivations during this period (Eriksen and Niemi 1981). When the Berlin Wall fell and the Communist regime in the Soviet Union collapsed, the mining industry there shut down as well (Viken and Nyseth 2009). This sort of state industrial support legitimized through military interests became history after the Cold War. Some of these cities play national roles now: Hammerfest is Norway's leading site for the development of natural gas from the Barents Sea, Kirkenes is the nation's gateway towards Russia, and Tromsø has several national roles, including as a university city and as the Arctic capital.

Some Arctic cities are quite new, less than a hundred years old, and came into existence as a result of a mine, a harbor, or a fleet base. There are some older cities like Tromsø—more than 200 years old—which grew because of commercial trade and maritime traffic. Taking this into account, it can be said that many Arctic cities are not a consequence of self-sustainable growth to any high degree, which means that they are as fragile and as vulnerable as the resources and the politics that fueled their growth. Because these cities are relatively young in comparison to cities in central Europe, and because their development was often a result of industrial forms of production, their visual aesthetics are dominated by modern post-war architectural design. Few of these cities have buildings of historical value. Some of them, like Hammerfest and Kirkenes, were completely destroyed during World War II, and their current forms were shaped in the 1950s and 1960s. Tromsø, with its small wooden houses, is an exception in the Arctic, even though city growth, urban planning, and several city fires have destroyed much of the eighteenth and nineteenth-century architecture.

Towards an Understanding of the Arctic City: The New Metropolis?

This overview of northern Scandinavian urbanization processes has revealed the Arctic city's ambivalent position vis-à-vis a common understanding of urbanity: cities in the Scandinavian Arctic are distinct, yet comparable to urban development in other parts of the world. A closer look soon unveils small settlements and towns in transformative stages, within a network of relations that not only crosses the Arctic landscape and national borders, but is of global reach. The Scandinavian welfare state provides these communities with public institutions and infrastructure to fulfill their role as regional urban centers in the modern world. Ways of being urban and ways of making new kinds of urban futures are indeed diverse here (Robinson 2006; Hubbard 2006). "One size does not fit all," as Thrift writes about cities and modernity (Thrift 2000). Following Robinson's postcolonial perspective on urban studies, I strongly agree that theorizing about cities should be based on a greater diversity of urban experiences.

WORK CITED

Amin, Ash, and Nigel Thrift. 2002. *Cities. Reimagining the urban*. Cambridge: Polity Press.

Ascher, Francois. 2007. Multimobility, multispeed cities: A challenge for architects, town planners and politicians. *Places* 19(1): 36–41.

Bæck, Unn-Doris. 2004. The urban ethos: Locality and youth in North Norway. *Young* 12(2): 99–115.

Bæck, Unn-Doris Karlsen, and Gry Paulgaard (ed). 2012. *Rural futures? Finding one's place within changing labour markets*. Stamsund: Orkana Akademisk.

Bell, David, and Mark Jayne. 2009. Small cities? Towards a research agenda. *International Journal of Urban and Regional Research* 33(3): 683–699.

Benediktsson, Karl. 2009. The industrial imperative and second (hand) modernity. In *Place reinvention: Northern perspectives*, eds. Torill Nyseth and Arid Viken, 15–31. London: Ashgate.

Castells, Manuel. 1996. *The information age*, vol 1. Oxford: Blackwell.

Dybbroe, Susanne, Jens Dahl, and Ludger Müller-Wille. 2010. Dynamics of Arctic urbanization. *Acta Borealia* 27(2): 120–124.

Eriksen, Knut Einar, and Einar Niemi. 1981. *Den finske fare. Sikkerhetsproblemer og minoritetspolitikk i nord*. Oslo: Universitetsforlaget.

Granås, Brynhild. 2015. Fra stedsteori til teori om sammenkastethet: Materialitet, historie og geografi i lesninger av hundekjøring i Norge. In *Med sans for sted. Nyere teorier*, eds. Marit Aure, Jørn Cruickshank, Nina Gunnerud-Berg, and Britt Dale, 299–316. Bergen: Fagbokforlaget.

Guneriussen, Willy. 2008. Modernity re-enchanted: Making a 'magic' region. In *Mobility and place. Enacting Northern European peripheries*, eds. Jørgen Ole Bærenholdt and Brynhild Granås, 236–244. London: Ashgate.

———. 2012. *Arctification*. Unpublished lectures. Tromsø: UiT, The Arctic University of Norway.

Hinchcliffe, Steve, and Sara Whatmore. 2006. Living cities: Towards a politics of conviviality. *Science as Culture* 15(2): 123–138.

Hubbard, Phil. 2006. *City*. London: Routledge.

Kishigami, Nobuhiro, and Molly Lie, eds. 2008. *Inuit urbains/urban Inuit*. Special issue, *Études/Inuit/Studies* 32(1): 5–11

Munkejord, Mai Camilla. 2009. Reinventing rurality in the North. In *Place rein-vention: Northern perspectives*, eds. Torill Nyseth and Arvid Viken, 203–219. London: Ashgate.

Nyseth, Torill, and Arvid Viken, eds. 2009. *Place reinvention. Northern perspectives*. London: Ashgate.

Nyseth, Torill, and Paul Pedersen. 2014. Urban Sámi identities in Scandinavia: Hybridities, ambivalences and cultural innovation. *Acta Borealia* 31(2): 131–151.

Robinson, Jennifer. 2006. *Ordinary cities. Between modernity and development*. London: Routledge.

Sassen, Saskia. 1991. *The global city: New York, London Tokyo*. Princeton: Princeton University Press.

———. 2012. When the centre no longer holds. Cities as frontier zones. *Cities* 34: 67–70.

Thrift, Nigel. 2000. Not a straight line but a curve, or cities are not mirrors of modernity. In *City visions*, eds. David Bell and Azzedine Haddour, 233–251. Prentice Hall: Harlow.

Viken, Arvid. 2011. Naturbasert turisme i nord. Ytre påvirkning – lokal tilpasning. In *Hvor går Nord-Norge? Tidsbilder fra en landsdel i forandring*, eds. Svein Jentoft, Jens Ivar Nergård, and Kjell Arne Røvik, 175–188. Stamsund: Orkana.

Viken, Arvid and Bjarge Schwenke Fors. 2014. *Grenseliv*. Stamsund: Orkana.

Viken, Arvid, and Torill Nyseth. 2009. Kirkenes – A town for miners and minis-ters. In *Place reinvention. Northern perspectives*, eds. Torill Nyseth and Arvid Viken, 53–73. London: Ashgate.

Viken, Arvid, and Robert Petterson. 2007. Sámi perspectives on indigenous tour-ism in Northern Europe: Commerce or cultural development? In *Tourism and indigenous peoples*, eds. Richard Butler and Tomas Hinch, 176–187. London: Thompson.

Wirth, Louis. 2003 [1938]. Urbanism as a way of life. In *The city reader*, eds. Richard T. LeGates and Frederic Stout, 97–105. London: Routledge.

CHAPTER 5

Cod Society: The Technopolitics of Modern Greenland

Kristian H. Nielsen

The Arctic is one of the regions of the world where the relationship between hunting and fishing societies and the resources on which they depend is the most fragile. Varying degrees of availability of natural resources, accompanied by periodic fluctuations in climate, means that Arctic regions have seen prosperity and poverty follow each other in rapid succession. The history of cod fishing in Greenland is one of the more extreme examples of this fragility. Owing to warmer Arctic temperatures in the early part of the twentieth century, Greenland's cod fisheries expanded enormously at the expense of seal hunting, but a steep decline in the 1960s led to the subsequent transition from a cod-fishing to a shrimp-fishing economy (Hamilton et al. 2003). Both transitions, from seal to cod and then from cod to shrimp, were the result of the interplay between natural and social forces. These transformations of Arctic ecology and culture have taken place in gradual shifts, with one complex technopolitical assemblage slowly supplementing or surpassing another. This chapter traces the emergence of a "cod society" in Greenland after World War II, a society that consciously planned and assembled with cod fishery as the mainstay of the economy. In particular, this chapter will analyze the white paper on Greenland published in 1950 by the Greenland Commission, a group

K.H. Nielsen (✉)
Centre for Science Studies, Ny Munkegade 118, 8000 Aarhus C, Denmark

© The Author(s) 2017
L.-A. Körber et al. (eds.), *Arctic Environmental Modernities*,
DOI 10.1007/978-3-319-39116-8_5

of Danish government officials and a broad range of experts, in which cod fishing on an industrial scale was seen as one of the most important means of producing economic growth and social welfare in the country. Emphasizing the links established between natural resources, new technologies, and social change, I examine the unintended consequences of the Greenland Commission's grand modernization scheme.

Cod, in Mark Kurlansky's (1997) biography of "the fish that changed the world," is the symbol of a global crisis in the relation of humans to nature. He argues that cod has been the driving natural force behind wars and revolutions, that it has fed populations and formed the basis of whole economies, and that cod was one of the reasons why Europeans first crossed the Atlantic, providing the nutrition needed for their endeavors. Tracing the fall of cod stocks in the latter part of the twentieth century, Kurlansky notes that the introduction of industrial fishing techniques led to overfishing with dire ecological and socioeconomic consequences. While Kurlansky sees cod as a prime mover of social development, and socio-technical developments like industrial cod fisheries as being the main cause of the decline in cod stocks, I want to stress that Greenland's cod society was built on hybrid forms of power embedded in natural resources, technological infrastructure, and legal-administrative systems. Cod can certainly be seen as the driving natural force behind the political reforms in Greenland from 1950 onwards; yet, its power had to be negotiated in relation to highly diverse forces such as Danish sovereignty over Greenland, the construction of technological infrastructure in Arctic environments, and the enactment of a more independent political administration in a sparsely populated area. To complement Kurlansky's narrative of decline, I propose a historical one that emphasizes the shifts and displacements in the interplay between cod and society. In contrast to Kurlansky, however, whose work is fundamentally an elegy to a lost hunter society, I want to avoid nostalgia by stressing that unintended consequences, or overflows, do not necessarily lead to degradation, but have the potential to give rise to new forms of expertise and new forms of technopolitical assemblages.

Timothy Mitchell's work is helpful for understanding the relationship between nature, technology, society, and politics. Though he focuses on the construction of the modern Egyptian state and contemporary "carbon democracies," his theory as explained in *Rule of Experts* (2002) also pertains to the modernization process of Greenland. Mitchell explores the way natural entities such as the malaria parasite interacted with governmental practices, financial exchanges, and new methods of calculation and circulation to produce what he calls the "technopolitics" of modern Egypt. Hybrid forms of expertise were needed to produce new knowledge about the many socio-technical projects, like the Aswan Dam, that were proposed to make use of nature and promote modern structures in Egyptian society. Mitchell emphasizes that the engineering expertise proved insufficient to contain natural and social forces. For example, the construction of the Aswan Dam enabled the malaria mosquito to spread, causing new epidemics that threatened to destabilize the new technopolitical order, and which led to the development of new forms of expertise in public health.

The influential work *Carbon Democracy* (Mitchell 2011) also treats natural, technical, and social forces on a par. Mitchell sees close connections between the rise of democratic movements and the mining of coal, which provided the power source essential to the first industrial societies. By acquiring the ability to shut down coal production or in other ways regulate the flow of carbon, for example by means of sabotage, miners in the latter part of the nineteenth century attained a position of power and were able to force governments and industry leaders to listen to their demands. The co-assembly of a coal-based industrial society and mass democracy led to significant changes in the mindset of workers, managers, and politicians. With the increased usage of oil in the course of the twentieth century, the flows of carbon became much more difficult for workers to interrupt and easier for companies and governments to control. Producing oil requires a smaller workforce, and the distribution of oil by means of pipelines, trains, and tank ships can be done with relatively little human labor. Moreover, because of the fluidity and lightness of oil, it can be shipped across oceans. By geological accident, the largest oil deposits were found far from the established industrial centers and in politically unstable regions, in particular in the Middle East, which allowed for new machineries of control: outsourcing manufacturing overseas to countries with lower paid and less unionized workers threatened the latter in the industrialized West with lower wages and unemployment, and also tied the flows of oil to the US dollar and recycling payments for oil into arms purchases, thus fueling political instability and minimizing local political control over oil production.

The notion of technopolitics captures the co-construction of political power and technological systems. The concept not only denotes the strategic use of technology to enact political goals, but also alerts us to the unpredictable effects of technology and politics. Cod fisheries, made viable due to climatic changes in the early part of the twentieth century, would enable Greenland to utilize one of its few, abundant resources for purposes of socio-economic development. In contrast to seal hunting, which was limited to coastal areas and only able to sustain a limited number of people dispersed along the coastline, commercial cod fishery significantly enhanced primary production per inhabitant, making growth and development possible in the country. The mobilization of a sea-going fishing fleet supplemented by large-scale processing facilities on land specifically enabled the kind of modernity described by concepts like concentration and development, as advocated by the Danish and Greenlandic reformers. However, the case of cod technopolitics in Greenland also makes clear the recalcitrance of both human and non-human actors in conforming to novel configurations of modernist planning and technical expertise.

Technopolitics of Cod

Around 1920, warmer currents brought Atlantic cod and other fish to West Greenland waters. At the same time, sealing, the traditional livelihood of Greenlanders, declined rapidly due partly to this climatic change and partly to overfishing. Greenland was under Danish colonial rule at that time and was closed to outsiders, except for visitors carrying out certified scientific activities. Trading in Greenland was regulated by a Danish state enterprise and was for the most part based on a system of barter. Although some Danish civil servants, expressing the same kind of nostalgia for hunting as Kurlansky, argued in 1918 that "the natives are happier if they keep to their centuries-old occupation—seal hunting," Greenland's cod fishery continued to expand throughout the 1920s, reaching a small peak in 1930 (Mattox 1973, 116). Prior to 1950, cod fishing had remained relatively simple. Using small rowboats or dories, and hand or long lines, and operating more or less exclusively in the summer months, Greenlandic fishermen landed the fish for sale at the designated sites of the trade monopoly, where fresh cod could be salted and dried.

Cod fishing thus began as a seasonal and subsidiary activity for Greenlanders. By the outbreak of World War II, however, it had become the main occupation in the south of the country. This new-found independence for many Greenlanders strengthened political movements calling for greater self-determination. The war years had also provided Greenlanders with new perspectives on the outside world due to the fact that it had been controlled by US forces under an agreement with the exiled Danish government in Washington DC, while Denmark was occupied by Nazi Germany. Radio had been introduced in 1925, and the broadcast service was expanding, primarily due to the editors of the new newspaper in Danish, *Grønlandsposten*. The establishment of US military bases in Greenland during the war meant that Greenlanders were exposed to Western music and images of living conditions in the USA, some for the first time ever (Beukel et al. 2010; Nielsen 2013; Sørensen 2006).

After the war, Greenland and Denmark re-established political ties, but for many reasons it had become impracticable to continue as before. The international emphasis on decolonization, coupled with the new-found aspiration for self-determination in Greenland, required redefining Danish presence in the country. At the same time, the USA was seeking to expand its military presence in Greenland and offered to buy the island from Denmark in 1946. The Danish government rejected the offer, well aware that it had to take a more active stance. In order to enable a new postcolonial regime in Greenland, while also affirming Denmark's sovereignty over it, there was a perceived need to embark on a long-term and planned process of socio-technical development. In addition, Danish fishermen and fishing associations were calling for an opening up of trade in Greenland, which had been monopolized by the Royal Greenlandic Trading Company (Den Kongelige Grønlandske Handel) (Beukel et al. 2010; Sørensen 2006).

Following negotiations between leading Greenlanders and the Danish authorities, it was agreed that Denmark should initiate reforms aimed at modernizing Greenland. To this end, the Greenland Commission (Grønlandskommissionen) was established on 29 November 1948 "to examine the problems faced by Greenland with respect to its social, socio-economic, political, cultural, and administrative development and, on this basis, submit a report setting out proposals for future guidelines concerning these issues" (Grønlandskommissionen 1950, vol. 1, 5). The main Greenland Commission had 16 members, including four representatives of Greenland's two provincial councils. Nine sub-commissions were set up, consisting of a total of 105 members, only 12 of which were Greenlanders. At the recommendation of the Commission, Greenland was to switch from a barter to a money economy, opening up its markets to foreign investment and international trade. Gradually, the modern infrastructures of health, education, industry, and political administration were to be introduced. When the Greenland Commission took up the question of how to respond to the country's future challenges, cod fishery was seen as the most important driving force of the new economy:

> Cod fishing, according to the Commission, will be the main occupation in Greenland in the future, so if Greenlandic society by its own means is going to maintain the current standard of living, and possibly improve it, the primary production per inhabitant has to be increased. It will be one of the most important tasks of the Commission to suggest steps to increase the catch of fish. (Grønlandskommissionen 1950, vol. 1, 83)

The final report of the Commission is in many ways a remarkable document, expressing not only the high-modernist development ideology globally in vogue at the time, but also Danish sensitivities to the specificity of the situation in Greenland, cultivated by many years of colonial rule. The report comprises a total vision for modern Greenland, based on the premise that Greenlanders needed help to enter the modern age, and with due concern for the special environmental and cultural conditions in the country. The harsh Arctic climate and the extensive, sparsely populated territory—approximately 22,000 people lived in Greenland in 1950—meant that new technical and administrative infrastructures would have to be designed especially for the country. Danish expertise would be useful, but it would have to be transformed in order to become applicable in a Greenlandic context. The fact that Greenlanders had their own language, history, and cultural habits implied that the new technopolitics enforced in the country would in no way aim for a complete alignment between it and Denmark. It was emphasized again and again that Greenland's distinctive character would have to remain intact, even if the process of modernizing it would necessarily result in significant structural changes (Grønlandskommissionen 1950, vol. 1).

In its report, the Commission stressed that, due to the ongoing changes in the living and occupational conditions of the Greenlanders—from seal hunting and a barter economy to cod fishing and a money economy—the time was right to include Greenlanders as equal members of Danish society, while also opening up economic and cultural relations between Greenland and the rest of the world. Until now, most historical interpretations of the Commission's report have concentrated on its political consequences for Greenland and Denmark, but I am more concerned with the report's many-sided concerns with technology, natural resources, political administration, and socioeconomic development. The report particularly emphasized that the Commission's vision for modern Greenland was a technopolitics of cod. It built on the premise that the country's emerging economy could be supported by industrial cod fisheries centralized in larger towns, and that developments in other spheres of society such as health, education, culture, religion, administration, and technological infrastructure had to be planned accordingly.

The technopolitics of cod represents the Danish and Greenlandic reformers' attempts at building a modern welfare state in Greenland based on reinforced and redefined political ties to Denmark and the new income provided by cod fisheries. Modern development in Greenland was seen as dependent on the new cod fishing fleet with its advanced technical facilities and competent fishermen. This again necessitated the emergence of a modern welfare state with educated and disciplined citizens, concentrated in larger settlements where education, administration, and cultural life would be able to thrive, which would allow for enough cod to be landed and processed in the country. Facilitating the movement of cod was thus essential to the mobilization of public aid. The technopolitics of cod implied that new fishing technologies would be put to use for the benefit of the population, but also required that the latter move accordingly in demographic and educational terms.

CONCENTRATING AND DEVELOPING GREENLAND

Sealing in Greenland necessitated a strong dispersal of the population; a larger population assembled in one area, rich in seals, necessarily would have led to over-exploitation of the seals and led the seals to seek for quieter areas. … In other words, the dispersed population in previous times not only was advantageous, but absolutely necessary. … The dispersed settlements, however, are not necessary for its inhabitants to work as fishermen, on the contrary. For the fisherman, it is first of all vital to live close to rich fishing grounds in order to avoid long and time-consuming transportation back and forth. Secondly, he has to live close to places where he can sell his catch and buy goods for his income. … The structural changes in Greenland, induced by the transition from sealing to fishing, favor or rather necessitate the concentration of the Greenlandic population in larger and much fewer settlements. (Grønlandskommissionen 1950, vol. 1, 22–23)

Concentration and development were key words in the technopolitics of cod described in the Commission's report. Allowing for sufficient amounts of cod to flow from the waters off West Greenland to international markets required first of all the introduction of new modern fishing vessels to increase the primary production. This entailed a shift from the smaller row or motor boats most commonly used to larger cutters of approximately ten to fifteen tons. The number of cutters needed would be smaller than the number of motor boats, and the crew on such cutters would be larger. Due partly to highly beneficial loans from the Danish government, motor boats were easily obtained by individual fishermen, but the more expensive cutters would have to be financed through stricter agreements. Reducing the number of fishing vessels by increasing their size, while also increasing the sophistication of the technical equipment aboard each vessel, was seen as a way to make cod fishing more efficient (Grønlandskommissionen 1950, vol. 5, 77–102).

Along with the concentration and development of fishing vessels and fishermen, the Commission proposed changing the processing method of cod from drying and salting to quick-freezing. Following World War II, Greenland had about a hundred smaller fishing houses where cod (and other catches) could be dried or salted. The drying process took about three months and most often had to be carried out in the open air due to lack of indoor space. Moreover, the production of salted fish was vulnerable to temperatures below 0 °C. Despite these difficulties, Greenlandic cod was considered to be of high quality and was exported primarily to Italy, Greece, Egypt, Spain, and Portugal. Quick or flash-freezing was developed in the 1930s by the American inventor Clarence Birdseye, who wanted to make frozen fish available to people living far from coastal areas that "would be in every way as desirable as fresh" (Hilder 1930). The first quick-freezing plant was constructed in Tovkussak by the Danish-owned Greenland Fishing Company. The introduction of additional quick-freezing plants at the most profitable landing sites would make the processing of cod possible in all weather conditions and all year round. Although the Commission argued that in the near future quick-freezing would have to supplement, but not replace, drying and salting, ultimately the number of landing sites was reduced while the amount of cod processed at each site was increased and the techniques of cod processing were refined (Grønlandskommissionen 1950, vol. 5., 110–116).

Implementing technological improvements and education in Greenland's fishery, the Commission predicted that the flow of cod from the sea would not only be concentrated, but also expanded. The same predictions were made in relation to the flow of people in Greenland. The population issue was in many ways more delicate than the cod issue.

Whereas cod in theory could be directly controlled by means of advanced fishing equipment —barring any climatic changes, as the Commission was well aware—the population of Greenland had to be dealt with in an indirect manner. The Commission noted that ongoing demographic changes in the country already followed the pattern of concentration and expansion. The population was increasing by 2 % annually—twice the rate of increase in the rest of Denmark—and Greenlanders were moving away from the northern and southern parts of the country to larger settlements or colonies situated in the central part of the west coast. This had to do with the decline of traditional seal hunting and the rise of cod fishing, due partly to Arctic warming and partly to the overhunting of seals in the Northern Atlantic. The population was drawn towards the rising fishing trade and to other means of employment related to the exploitation of natural resources in Greenland, such as coal mining in Kutdligssat. Unsurprisingly, this spontaneous concentration was seen as a positive thing, not only because of the new possibilities within fishery and trade, but also with respect to public health, education, administration, culture, and the economy—in fact all of the issues addressed by the various sub-commissions of the Commission (Grønlandskommissionen 1950, vol. 1., 21–27).

Regarding some of the smallest and poorest settlements, the Commission raised the question whether relocating the population against its own will would be necessary, given that people in the most remote areas would have great difficulties in sustaining themselves. The answer was categorically negative, and therefore other less coercive means, such as information campaigns, were proposed:

> It has always been clear to the Commission that relocation of the population by force has to be disregarded as being in opposition to the liberal and democratic principles cherished by Danish society and to the spirit of freedom and individualism valued by Greenlanders. Yet, it is necessary in writing and speech (radio) to make the Greenlandic population aware that there are good reasons for concentrating the population in larger settlements where the occupational and commercial opportunities are good. Moreover, the Commission believes that the population itself will become interested in relocating to the most suitable places, when proper occupations and trades, good schools, hospital services and doctors, and other cultural goods, in particular healthy housing, will make it attractive to do so. (Grønlandskommissionen 1950, vol. 1, 28)

In 1953, the liberal and democratic principles emphasized by the Commission were overridden when the government, in order to secure the extension of the US Thule Air Base in agreement with American security policies, forced 27 Inughuit families—116 people—to relocate from their home in Uummannaq to Qaanaaq, approximately 150 kilometers to the north. At the time, the relocation was described as voluntary and in agreement with the policy mentioned in the quote above. The Inughuits were offered newly built houses and compensation in the form of goods and equipment from the local trading post. The relocation took place just one month before Greenland's formal status was changed from colony to Danish county. However, investigations undertaken in 1996 in relation to compensation claims concluded that the relocation had been forced (Walsøe 2003).

INCIPIENT URBANIZATION

> The cultural, political and economic maturation of the Greenland population, which is the end goal for the Danish work in Greenland, is inhibited by the present low standards of living. It is not putting things too strongly to say that improvements in the standards of living in Greenland is one of the preconditions for obtaining the full effect of the propositions of the Commission in health, cultural, political, and economic areas. In particular, one should not underestimate the effects that the poor living conditions, directly or indirectly, have had in terms of the current low economic efficiency found in Greenland. (Grønlandskommissionen 1950, vol. 4, 51)

Scholars have examined how, during the middle decades of the twentieth century, development projects and planning in the Arctic, Asia, Africa, and elsewhere have been driven by ideas of high modernism and technological progress (Engerman et al. 2003; Cullather 2010). The cases of two Canadian Arctic towns, Iqaluit (then known as Frobisher Bay) and Inuvik, show how federal officials advanced elaborate plans for urban development in order to draw native Northerners into conditions of modern living. Like Greenland, the Canadian Arctic was undergoing what the authors Farish and Lackenbauer call "incipient urbanization" (2009, 539). (For discourses of Arctic urbanism, with a focus on mainland Scandinavia, see chapter "Arctic Urbanization: Modernity Without Cities"). This shift also took place in Greenland, from a semi-nomadic lifestyle or living together in very small communities to the concentration of the population in larger towns, which were increasingly equipped like Danish towns, with water supplies, renovation work, and modern housing, albeit with due consideration given to the special conditions found in Greenland.

The Greenland Commission observed that, compared to Danish standards, the living conditions in Greenland generally were poor, if not deplorable. Most houses had just one room, and worse, were cold, damp, and draughty. The lowly housing standard resulted in health problems and cultural degradation, since it was impossible to read and write in the dark, freezing, and cramped rooms. Many Greenlanders suffered from rheumatism and tuberculosis. Taking into account that the average number of occupants per house was about six, the Commission therefore recommended introducing standard houses of variable sizes. Most Greenlanders were used to building their own houses, but this tradition had to be discontinued as it was part of the reason why so many of them were deficient and lacked adequate facilities. In order to carry out this plan, Danish building expertise needed to be introduced, and loans from the Danish Government offered on favorable terms (Grønlandskommissionen 1950, vol. 4, 44–78).

The first urban planners to visit Greenland in 1950 wholeheartedly supported the technopolitics of concentration and incipient urbanization. Agreeing that the development of the cod industry in fewer, but larger units had to be accompanied by similar developments in towns, the planners argued that urban development had to proceed according to local conditions. Where the Commission had suggested wood as the least problematic construction material, the architects advocated concrete as the building material of the future. The development of the cod industry should lead to a "more concentrated form of settlement" than the one seen in Greenland at the time, it was argued, and concrete was best suited for this purpose (Andersen 1951, 36). The urban planners wanted to go a step further than the Commission, which had suggested local planning as one of the requirements for incipient urbanization. The planners argued that, since all of the small, more or less independent communities in Greenland now would be connected together in "one Greenlandic society," the construction of harbors, a fishing industry, housing, schools, hospitals, cinemas, and even bakeries no longer pertained merely to local affairs (Andersen 1951, 113).

The most conspicuous example of modern urban planning was the construction of Blok P in Godthaab (Nuuk) (see Fig. 5.1), which was inspired by the functionalist architectural style in vogue at the time in

Fig. 5.1 Blok P in Nuuk, built by the Technical Organization of Greenland, a Danish governmental agency (Photo: Gunnar P. Rosendahl). Reprinted with permission of the Greenland National Archives

Denmark. At the time of completion in 1967, it was the largest estate in the Kingdom of Denmark. The building was five stories high with 64 apartments on each story. It accommodated around 1 % of the entire population of Greenland. Staying close together in apartment blocks was very different to traditional living in the form of smaller and dispersed settlements. Some Greenlanders did not feel comfortable living in Blok P, while others treasured the modern facilities such as running water and soon found ways to integrate traditional ways of life into the small apartments, by using the balconies for drying meat and curing skin for example. Criticism began to emerge as the first wave of enthusiasm for Blok P abated, and as the estate slowly turned from avant-garde housing to a slum. Some saw Blok P as material evidence of the Danish attempt to recolonize Greenland by means of modern amenities. Others were appalled at the stark aesthetic contrast between Arctic nature and these urban "living machines." In 2012, the local authorities demolished Blok P (Hilker and Diemer 2013).

CONFIGURATIONS OF CHANGE

It is tempting to see Danish endeavors at modernizing Greenland as the collision of two worlds. On the one side there are the Danish reformers who, although always careful to emphasize that the modernization process was being initiated for Greenland's own good, launched a grand development program. Subscribing more or less to the then contemporary ideals of modernist planning, they basically wanted to transform Greenland into a modern welfare society, complete with technical and administrative infrastructure, with well-educated, healthy, and highly disciplined industrial citizens. On the other side there is Greenland's traditional culture based on small-scale, highly autonomous communities with relatively few connections to the outside world, who were totally unaccustomed to the pace and complexity of modern living. This view was implicit in reports on topics ranging from water drainage to legal affairs authored by Danish experts, who visited Greenland to collect information to be used in the Commission's report during the summers of 1948 and 1949. With a few exceptions, they described conditions in Greenland as "appalling" from a Danish perspective. The notion of a cultural collision gained additional support in the 1950s when Danish social scientists began taking an interest in the consequences of modernization for Greenland's population, discovering that the social ills that accompany modern society were also found in Greenland, and with added strength due to the collision of the two worlds (for a discussion of the recent Greenlandic reconciliation commission as a means of coping with this collision, see chapter "The Greenlandic Reconciliation Commission: Ethnonationalism, Arctic Resources, and Post-Colonial Identity").

The two-worlds colliding discourse is also strong in narratives about the interaction between humans and the environment. It can be seen in Kurlansky's work, and it is implicit in simple observations about the overfishing of seals, and then later cod. I agree that such narratives have merits, in part because they enable simple solutions to complex problems. If "traditional" ways of life suffer due to the impact of modern culture, then we have to protect them or at least lessen the consequences of modernity for them. If natural resources are depleted due to the impact of industrial fishing, then we have to protect them.

The technopolitics of cod, as envisaged by the Greenland Commission, was seen as a way to bring welfare to Greenland, to connect it economically and socially to the rest of the world, and to reinvent the relation-

84   K.H. NIELSEN

ship between Denmark and it in a new postcolonial context. Using the natural environment, in this case cod, as the basis of a historical narrative about change in Greenland, I have explored another way of framing such complex relationships, namely as the shift from one technopolitical configuration to another. The emerging "cod society" of Greenland in effect resembled past Greenlandic societies in that it employed certain kinds of technology to structure relationships between the country and the rest of the world (Denmark, in particular), and between humans and nature. The difference between one and the other, I would argue, is ultimately not an essential one, as the metaphor of colliding cultures seems to suggest, but rather a difference in scale and timing.

The Commission's report carefully scrutinized the existing "seal technopolitics" of Greenland in order to find ways in which to imagine new forms of "cod technopolitics." Cod was the envisaged intermediary that would enable connections to be made between existing technopolitical configurations and future ones. Since cod fisheries were already well established in Greenland, it provided the Danish reformers with a "hook" that not only reached into the future but also extended into the immediate past. The total vision of Greenland as a cod society, produced by the Commission, consisted of a messy integration of existing and imagined natural and cultural factors.

WORK CITED

Andersen, Hugo Lund. 1951. *Byplanforslag i Vestgrønland: Narssaq, Sukkertoppen, Egedesminde, Godthaab*. Copenhagen: Greenland Department, Ministry of the State of Denmark.

Beukel, Erik, Frede P. Jensen, and Jens Elo Rytter. 2010. *Phasing out the colonial status of Greenland, 1945–54: A historical study*. Copenhagen: Museum Tusculanum Press.

Cullather, Nick. 2010. *The hungry World: America's Cold War battle against poverty in Asia*. Cambridge, MA: Harvard University Press.

Engerman, David C., Nils Gilman, Mark H. Haefele, and Michael E. Latham, eds. 2003. *Staging growth: Modernization, development, and the global cold war, culture, politics, and the cold war*. Amherst: University of Massachusetts Press.

Farish, Matthew, and P. Whitney Lackenbauer. 2009. High modernism in the Arctic: Planning Frobisher Bay and Inuvik. *Journal of Historical Geography* 35(3): 517–544.

Grønlandskommissionen. 1950. *Grønlandskommissionens betænkning*, 9 vols. Copenhagen.

Hamilton, Lawrence C., Benjamin C. Brown, and Rasmus O. Rasmussen. 2003. West Greenland's Cod-to-Shrimp transition: Local dimensions of climatic change. *Arctic* 56(3): 271–282.

Hilder, John Chapman. 1930. "Quick-frozen Food exactly like fresh." *Popular Science Monthly*, September, 26–27.

Hilker, Martin, and Rikke Diemer, eds. 2013. *Blok P: En Boligblok i Nuuk*. Copenhagen: Nordatlantens Brygge.

Kjær Sørensen, Axel. 2006. *Denmark-Greenland in the twentieth century*. Copenhagen: Museum Tusculanum Press.

Kurlansky, Mark. 1997. *Cod: A biography of the fish that changed the world*. New York: Walker and Co.

Mattox, William G. 1973. *Fishing in West Greenland 1910–1966: The development of a new native industry*. Meddelelser om Grønland 197. Copenhagen: C. A. Reitzel.

Mitchell, Timothy. 2002. *Rule of experts: Egypt, techno-politics, modernity*. Berkeley: University of California Press.

———. 2011. *Carbon democracy: political power in the age of oil*. London: Verso.

Nielsen, Kristian H. 2013. Transforming Greenland: Imperial formations in the cold war. *New Global Studies* 7(2): 129–154. doi:10.1515/ngs-2013-013.

Walsøe, Per. 2003. *Goodbye Thule: The compulsory relocation in 1953*. Copenhagen: Tiderne Skifter.

CHAPTER 6

Re-reading Knut Hamsun in Collaboration with Place in Lule Sámi Nordlándda

Kikki Jernsletten and Troy Storfjell

For many the "Arctic" is an exotic geographic abstraction, something different and fascinating, something somewhere "out there." For some of us, however, that abstraction is a difficult one, since our own well-known and well-loved corner of the Arctic is simply "home"—eminently familiar, quite specific, and not readily reducible to an abstraction of otherness. It is a Sámi place, our place, the place that makes us, as Sámi, who we are. And our relationship to the landscape, plants, animals, people, and other beings of our place goes back many generations, millennia even; for us this relationship includes the ghosts and stories of those who once walked, and still inhabit, our place. Place makes us who we are because we are, in fact, produced by its complex networks of relationships within which we are situated. Like so many Indigenous peoples, we recognize that we *are* our relationships (Wilson 2008, 69–79).

In this chapter we explore some of the ways that Indigenous Sámi understandings of place and time produce our own systems of knowledge about our specific portion of the "Arctic," meaning that portion with which we have a specific relationship. Although these knowledge systems can differ

K. Jernsletten (✉)
Independent Scholar, Harstad, Norway

T. Storfjell
Pacific Lutheran University, Tacoma, WA 98447, USA

© The Author(s) 2017
L.-A. Körber et al. (eds.), *Arctic Environmental Modernities*,
DOI 10.1007/978-3-319-39116-8_6

significantly from traditional academic ways of knowing, we believe that our epistemes and intellectual traditions can also inhabit the university, and here we join other Indigenous scholars from around the world in working to make a place for Indigenous knowledges in academic space. In this case we direct a place-based Sámi gaze at a well-known piece of "Arctic" literature, written by one of Norway's most canonical and controversial authors, Knut Hamsun (1859–1952). We take a look at how *Growth of the Soil* (*Markens grøde*, 1917) attempts to erase Sámi presence in its fictionalized treatment of an interior landscape in northern Norway, acknowledging that the specific place where he wrote the novel, and whose traces are clearly evident in it, is in fact a place rich in Sámi history and continued presence. We look at how Hamsun's discursive intervention into the local Sámi communities of Divtasvuodna (Tysfjord), Hábmer (Hamarøy), and Stájgo (Steigen), in the county of Nordlándda (Nordland), has affected some of the Lule Sámi living there, serving as a well-known articulation of the colonial Norwegian combination of disdain and disregard with which they have had to contend for several generations. But we refuse simply to accept the narrative of Sámi as victims. Instead we also investigate the ways that this Sámi place has produced Hamsun himself, and the strong traces that the Sámi presence here produces in the novel. In response to a century of scholarship that all but completely ignores the Sámi and our relationship to *Growth of the Soil*, we make the Sámi place, including its voices and stories, central in our analysis of the work and its author. In doing this, we argue that a place-based approach grounded in Indigenous methodologies can contribute not only to Sámi understandings of Hamsun and Lule Sábme (Lule Sámi Land), but also to a general understanding of Hamsun and his novel.

If, as Jacques Derrida has so famously suggested, every text contains its own methodology (2005, 199–201), then it follows that every place does so as well. Place is, after all, a text of its own, a site where richly layered significations and intertextualities overlap and interact. And yes, it might be tempting to see this sort of definition as a kind of colonization of the non-written by the culture of literacy—since it's doubtful that any oral cultures would use a term for a piece of writing to signify the idea of place. But perhaps we can let this slide for the moment and move forward with the analogy. We have chosen to write this, after all, and so have already agreed to enter into the realm of literate culture. It's a compromise we feel is necessary not only to reach the wider, non-Sámi academy, but also to converse with many of our Sámi colleagues, yet another instance of that time-honored Sámi tradition of speaking in at least two different registers at the same time (cf. Gaski 1993).

This is exactly the sort of balancing act that Indigenous academics find ourselves having to perform quite often, though. Geographer Kali Fermantez (Kanaka Maoli), for instance, writes that native scholars need to rely on our shape-shifting abilities to mediate between the worldviews and demands of our home communities and those of the academy, donning different forms for conversing with each group (2013, 103). In saying this he echoes a number of other contemporary Indigenous scholars (cf. Johnson 2013, 128; Kuokkanen 2007, 51–54; Smith 1999, 37–38, 69; Acoose 1995). After all, even though we are Indigenous, we are also scholars; we have been educated within the colonizing space of the university, and have been disciplined to one degree or another by its institutional structures and traditions. The important thing for those of us working with Indigenous methodologies is to open up that university space that we already inhabit in order to make room for Indigenous ways of knowing. We are working to create Indigenous spaces—and Indigenous places— within the academy, alongside the traditional disciplinary structures and approaches that, though generally passed off as universal, actually derive from a very specific intellectual moment in nineteenth-century Germany (Anderson 2010).

INDIGENOUS METHODOLOGIES

Our personal encounters with the cultural specificity of the academy have prompted a number of Indigenous scholars to challenge the ethnocentrism embodied in its traditions and ways of knowing, and since the turn of the millennium the emergence of Indigenous methodologies has provided a growing platform from which to develop ways of producing knowledge in the academy that are compatible with our cultures and their intellectual and philosophical traditions. In working to make room in the university for Indigenous epistemes and paradigms we are not seeking to supplant completely those older ways of doing things. Yet even the simple act of calling on the university to open itself up to more than one culturally specific set of traditions and lore does have far-reaching repercussions. Merely asking that the university recognize its dominant epistemic and ontological frameworks as culturally specific, and neither universal nor the only valid way of producing knowledge, is to call for a fundamental transformation in the university, or what Rauna Kuokkanen (Sámi) describes as "an opening up to a new way of seeing and conceptualizing knowledge as well as our relationships and responsibilities in terms of other individuals, groups, and epistemes" (2007, 159).

Kuokkanen is not the only Sámi scholar calling on the university to open itself up to Indigenous epistemes and knowledges. Central among the growing number engaging in this work is Harald Gaski, who has for many years written of the importance of the scholar's personal relationship to the material she or he studies, while also consistently including not only Sámi concepts and stories, but also traditional Sámi aesthetics, including such things as digression, humor, and the deliberate "layering" of communication in such a way that different messages are communicated to outsiders, to those with a superficial or partial familiarity with the community, and to those with true insider knowledge (cf. Gaski 1993, 2015). This Sámi aesthetic of refusing concise and totalizing statements in favor of a certain degree of deliberate ambivalence and a multilayered communication has also been treated as a tactic for Sámi scholarship by Kikki, in her dissertation (2012), and even Troy has tried once or twice (2016). Part of the point is that there is a certain subversive freedom in remaining less than fully restrained by rigidly transparent language. And part of it is that our knowledge is dialogic in nature, that it emerges from an ongoing conversation between people, and with place, story, and non-human beings. It cannot be completely pinned down, and that realization can itself bring some important insights to the academy.

While Indigenous methodologies are still relatively young, a consensus has clearly emerged. Practitioners agree that Indigenist research requires community involvement in designing projects and research questions and in implementing them; furthermore, Indigenist projects need to be relevant to these same communities, and their results should be shared with the communities that helped produce them. The ways in which such research projects are carried out must also respect community norms and sensibilities and follow local cultural protocols. They should be collaborative and grounded in an ethics of "relational accountability" (Wilson 2008, 7), or the idea that we are all situated within networks of relationships, and that we recognize that each of these relationships brings responsibilities with it. We are responsible to the places, people, and other living things with which we have relationships (Kuokkanen 2007, 44–45; Smith 1999, 10, 173, 176–77; Wilson 2008, 80–96).

Significantly, Indigenous methodologies also reject the traditional Eurocentric academic understanding of the mind as separate from a world that it perceives as object. Rather, the person-mind is part of the process of the world, which it can only observe from an embedded, insider's perspective (Kuokkanen 2007, 60; Ingold 2000, 101, 108). Indigenous methodologies not only see objectivity as an unobtainable ideological construct, they also see it as directly harmful and potentially dehumanizing. As Jeannette Armstrong (Okanagan Syilx) explains, "I try not to be 'objective' about anything. I fear those who are unemotional, and I solicit emotional response whenever I can. I do not stand silently by. I stand with you against the disorder" (2012, 40).

UNDERSTANDING *THIS* PLACE

> Place speaks to the holistic totality of human and nonhuman relations situated in a particular locale or region. Practitioners are recognizing that if field research is to be equitable, beneficial, and empowering, it must engage the complex realities and actors of place in a truly collaborative fashion. (Johnson and Larsen 2013, 8)

There is no other way to even begin to understand the place that we are looking at—the *kommuner* (townships) of Divtasvuodna-Hábmer-Stájgo—than to address the terrible history of local oppression and disparity, the competing and over-layering inscriptions of the space as Sámi and Norwegian, and of the Sámi as racially inferior. Inga Karlsen's autobiographical essay "En del av min oppvekst som same" ("Part of My Upbringing as a Sámi," 2011) helps to begin to explain some of the deep wounds that mark the psyches of many of the township's inhabitants. To be Lule Sámi from Divtasvuodna means, among many other things, to be an heir to this trauma. This is something that Lars Magne Andreassen's article "Skjult same i Narvik" ("Hidden Sámi in Narvik," 2011) also illustrates.

Of course, Norwegianization and racism are only part of the story of Divtasvuodna and neighboring municipalities. The trauma is there, to be sure, but there is more besides. There is also a deeply held love of place, an identification with specific locations in this fjord and mountain landscape that goes beyond the memories of one's own lifetime, the cloudberry bogs where one picked berries as a child, the mountain trails one has hiked with family and friends in the summer, or the frozen lakes one has skied across in winter on the way to a cabin in the woods (cf. Myrvoll 2010). Identification with specific locations in the area goes back several generations to the Sámi hamlets and villages where one's parents or grandparents lived, even if some of those places are no longer inhabited year round. A person is still from Vuodnabahta (Hellemo), for instance, even if he or she was born and raised somewhere else—be that Ájluokta (Drag), Gásluokta (Kjøpsvik), Hábmer, or even further away—as long as one belongs to one of the families that lived there back when it was still permanently settled, as long as one comes "back" in the summers and knows the people in the other summer homes.

One of the arenas in which this is readily apparent is in the Hellmokuppen, an annual football (soccer) tournament held in Måsske (Musken), in which Sámi teams based on various towns and villages along the Divtasvuodna fjord compete. Membership in these teams, and in the "Old-Timers" teams from the outer and inner fjord areas, is not necessarily based on where the player currently resides, but can often derive from which locale his or her family is known to belong to, or come from. This, of course, reflects a difference in temporal scale between Sámi and other Indigenous worldviews and those of mainstream, Western cultures. Although we're all familiar with the Western concept of time as linear and regulated by the clock, and with a rather short distance into the past to the horizon of relevance or immediacy, we also maintain familiarity with another sense of time, one that is non-linear, and in which the horizon of relevance and immediacy can be located much further in the past (cf. Bergman 2008, 19–20). We may "come from" a location in which we do not currently reside. Our families are from these places, and that means that we as individuals are too.

Similar temporal frameworks inform the way other Indigenous peoples (and Indigenous scholars) understand their relationships to place as well. As Fermantez comments, "in addition to spending a long time, even a lifetime in the community, temporal depth for Native scholars goes back generations. This, of course, speaks to genealogy and the knowledge that we gain from our ancestors" (2013, 112). As Indigenous researchers we have a broadened range of sources, owing to our own position in a complex network of relationships. The land, the trees, the lakes, streams, mountains, bogs, bays, inlets, islands, and the fjord itself are part of us. Or rather, we are part of these things; each of us is a person due to our relationships with them. And these things are not actually just *things*. They have agency. They have voices. They are part of a community that also includes animals and people. And the ghosts of the past.

Divtasvuodna, along with Hábmer and Stájgo, is part of the Julevsámegiella (Lule Sámi)-speaking portion of Sábme (Sámi Land), a cultural and linguistic strip that follows reindeer migratory routes from traditional summer grazing lands in Divtasvuodna and Hábmer over the mountains to the Swedish side and down through the hills and forests along the Julevädno (Lule) River, including places like Jåhkåmåhkke/ Dálvvadis (Jokkmokk) and Gállok (Kallak). Gállok is the place where, in the summer of 2013, Sámi protesters faced off against Swedish police and British-based Beowulf Mining, a transnational company working to turn prime winter grazing lands that have been used for centuries into an open pit mine for iron ore. Gállok, this culturally important place, is the place about which Beowulf CEO Clive Sinclaire-Puolton asked the rhetorical question "What local people?" characterizing it as *terra nullius*, empty land, in response to questions about how the locals feel about his mining plans (Tuorda 2014; saamicouncil 2012). Far from empty, Juvlev (Lule) Sábme, continuously inhabited since the last ice age, was its own cultural and linguistic unit long before it was carved up between the kingdoms of Norway and Sweden, whose border bisected it in 1751.

REACTING TO HAMSUN

An uninhabited, unpeopled landscape, though, is also the way that Hamsun portrayed the interior of this region in *Growth of the Soil*. This novel tells the story of the settler Isak, who builds a farm in the uninhabited wilderness of Nordlándda (Nordland) in a fictional setting that struck both of us as bearing an uncanny resemblance to the inland farm of Kråkmo, in Hábmer, where Hamsun rented a room and wrote a good portion of the novel.

In the story, Isak works tirelessly over the years to transform his first humble turf hut into a thriving farm and the cornerstone of a rural settler community. Yet at the beginning of this, the Nobel Prize-winning Hamsun's most popular novel, the land is presented as definitively empty:

That long, long path over the moors and into the forest, who has trodden it? Man, a human being, the first one who came here. There was no path before him. Later a few animals followed the faint tracks over the heaths and moors and made them clearer, and still later a few Lapps began to nose out the path and to use it when they were going from one mountain to another to see to their reindeer. This is how the path through the great common, the no-man's-land owned by no one, came into being. (2007, 4; for the Norwegian original, see Hamsun 1992, 145)

Of course, the real interior of northern Nordlándda was not uninhabited before the Norwegian settlers colonized it. It had been part of the reindeer herding lands of the Lule Sámi since the transition to large-scale herding some four centuries earlier, and for millennia before that it had been the site of their ancestors hunting and gathering ancestors, who had lived in the egalitarian, land-based communities known as siidat. This was Sábme: inhabited, claimed land, and even Hamsun's fictional treatment of it couldn't completely erase the Sámi inhabitation of it. Throughout the first half of the novel the Sámi are present in the textual wilderness, appearing as vagabonds, wandering beggars, and thieves, and being compared to maggots and vermin.

From a Sámi perspective, *Markens grøde* is a very painful novel to read, and the two of us have each published several critiques over the years in which we addressed the text's colonial and racist agendas (see Jernsletten 2004, 2006; Storfjell 2003, 2011a, 2011b). Our own reactions to reading it from a Sámi subjective position, and the hostility that many other Sámi readers have towards the book, made it important for us to problematize its oft-praised ideals of life lived in harmony with nature and expose their dark underbelly. This is a goal we shared from our student days in Romssa (Tromsø).

Yet even though documenting and analyzing the presence of colonial discourse in canonical national literature has its value, the critical tools we had employed could only take us so far, and we were each, on our own, working on new approaches to Hamsun and his work. In her doctoral dissertation Kikki had explored a Sámi poetics based on ságastallan (dialog), while also developing a collaborative, place-based methodology grounded in Sámi cultural practices that include asking permission from a place before building in it (2012). And during his sabbatical appointment as a guest researcher at Romssa universitehta (the University of Tromsø) Troy was exploring trans-Indigenous and Sámi-specific Indigenous methodologies. The two of us were brought together again when Troy was appointed to chair Kikki's dissertation *disputas* (defense) committee there. And in conversations following that festive event we decided to work together on a new, place-based analysis of *Growth of the Soil*, Knut Hamsun, and the Lule Sámi communities in the three neighboring municipalities of Divtasvuodna, Hábmer, and Stájgo.

We took our starting point in a set of Lule Sámi oral narratives about Hamsun that we had heard of via our mutual friend Trine Kalstad, from Hábmer. Trine's uncle, Nils Kalstad, had been a boy in Hábmer when Hamsun returned to live and work there as an adult, and had been interviewed by NRK (Norwegian National Broadcasting) back in the 1980s. One of the things that struck us as we listened to these recordings and read their transcripts was just how similar they were to Hamsun's writing, both in terms of language use and narrative technique. Kikki has already written about the novel's polyphonic inclusion of a Sámi counter-discourse (2004, 2006), but the clear witness that Uncle Nils's Norwegian use bore to the fact that Lule Sámi was his native language prompted us to wonder just how much Hamsun owed in other ways as well to the Sámi presence in his home municipality. How were his artistic style and themes shaped by the multi-ethnic, multilingual place where he grew up? And, equally importantly, of course, what is his effect on the Sámi community in that place today?

In Indigenous methodologies the situation of the researchers matters. Our relationships to the communities with which we work, and the responsibilities that those relationships confer, determine our work and need to be acknowledged. So it is important to point out that, while we are both Sámi, neither of us is Lule Sámi. Our ties are both to the North Sámi speaking part of Sábme—Kikki's to the River Sámi of the Deatnu (Tana) River, and Troy's to the Mark Sámi of southern Romsa (Troms) and Uffuohtá (Ofoten). And we both have ties beyond Sábme, too, to southern Norway and the United States, respectively. At the same time, though, our personal lives and relationships have involved each of us in the Divtasvuodna and Hábmer Lule Sámi communities for a number of years, meaning that we are not strangers there.

Our work on this project involved a good degree of collaboration. We met with the staff of the Hamsun Centre in Presteid, Hábmer, and the Árran Lule Sámi Centre in Ájluokta, Divtasvuodna, where we learned about Anna i Makkvatnet, Hamsun's Sámi housekeeper and hiking companion during his stays at the Kråkmo farm. At both centers we also benefited from a wealth of local knowledge and suggestions of people to talk to and places to visit, and our friendly and lively visit at Árran had the added benefit of helping us to shape our approach, questions, and goals in collaboration with local Sámi cultural experts and intellectuals. We stayed with Árran Director Lars Magne Andreassen, and with Trine Kalstad, who also acted as our ófelaš (pathfinder/guide) and photographer, and who pitched her lavvu (Sámi tent) and stayed with us at Strinda, her family's land across the lake from the Kråkmo farm. Beyond that we visited a number of homes in the area, talking to people about their relationship to the place, the effects of Norwegianization and racism on their lives, and their thoughts on Hamsun. And we got out and actually walked in the landscape, listening to it, collaborating with it, and asking its permission to build our own academic structure on it. A year later we returned to the community, gave a public presentation of our work to that point at Árran (and received a lot more useful feedback from local residents), and revisited key local places.

This methodology breaks with the traditional academic approaches to the discipline of literary study and criticism in which we were trained. For one thing, it assumes that real, extra-textual places matter in texts (even fictional ones), and that literary and scholarly representations of place invoke and participate in the network of relationships that make a place, and thus can be considered part of that place. We also value oral tradition, and the participation of the non-human residents of place. To the mainstream academy this can come as a shock. But it is quite in keeping with Indigenous ways of knowing, and has strong support in this emerging scholarly approach. Linda Tuhiwai Smith (Maori), for instance, notes that one of the key things that distinguishes Indigenous peoples from mainstream societies is our "spiritual relationships to the universe, to the landscape and to stones, rocks, insects and other things, seen and unseen," and that claims based on these relationships "have been difficult arguments for Western systems of knowledge to deal with or accept" (1999, 74). Jay T. Johnson (Delaware/Cherokee) and Soren C. Larsen claim that, in Indigenous research, "the importance of place as an active participant in collaboration cannot be underestimated. As a locale or situation where human and more-than-human others come together, the object of knowledge can be constructed, appreciated, and understood relative to its proper context and relationships" (2013, 14).

There is also support for the academic value of walking through the landscape from some non-Indigenous sources. Tim Ingold and Jo Lee Vergunst, for instance, point to its importance in building knowledge of place when they write that "movement ... is not adjunct to knowledge, as it is in the educational theory that underwrites classroom practice. Rather, the movement of walking is itself a way of knowing" (2008, 68). To which Johnson and Larsen add that place-based research "requires us to walk and dwell. This entails more than conventional participant observation, but rather an attunement to the embodied landscape as a primary way of coming to know ourselves in relation to others" (2013, 15).

ENCIRCLING HAMSUN

As we traverse the landscape, we are circling around Hamsun (much like the old-time Sámi hunters used to circle around the bear's den in the winter). But we're asking new questions—questions about the freedom that seems to open up for the Norwegian author in his confrontation with Sámi cultural space. We begin by looking at Sámi influence as a *necessary condition* for Hamsun's life and writing, for his epic, open, poetic language, his believable universe. We argue that the Sámi influence helped to shape this universe, that, though the author grew up in a nominally Norwegian reality, the Sámi presence was always there, lurking in the background, pointing the way towards another reality. It crept into the place's stories. Oral culture doesn't discriminate, after all, according to sources. Stories simply impose themselves, traveling and wanting to be told. Sámi influence is clearly present and prominent in the local storytelling scene, bringing with it its various beings, monsters, scandals, and magic. The child Hamsun, growing up in Hábmer, could not help but have been exposed to the living, oral, Sámi culture far to the north before the mass media had conquered the world.

Nature was also close and ever-present, just beyond the door, and the perception of it would have been shaped by the stories he heard. They came first. Growing up and into a culture, one also grows into a nexus of pre-formed ideas and ways of seeing and thinking. But, as Derrida explains, these systems of understanding are never complete; they all suppress those elements that challenge or undermine the structure, those pieces that have to be swept under the carpet in order to impose some sort of order on the chaos.

An open, perceptive mind can sometimes comprehend this, though, and artistic sensibilities often struggle to break down culturally constructed barriers and borders, the foreclosures that limit the world and our understanding of it. Many artists of modernity have felt that it was necessary to transgress these borders before anything truthful could be said about the human situation. For this sort of mind, on the lookout for ways to record the unconscious life of the mind, a radically different culture could conceivably function as a way out, an opening, or a door to the freedom between worlds—guovtti ilmmi gaskkas (between two heavens) (cf. Jernsletten 2012).

We have grounded our approach in Sámi communicative patterns, themselves the products of an active, living, storytelling tradition. The lyrical is alive, and can be found in children's stories, in riddles, words, and expressions that encompass human experiences and, in day-to-day use, perpetuate a people's values. It is found in the yoik, in the belief in the oral tradition's ability to contain hope, life, and the law. (Yoik is a traditional Sámi vocal music form in which the performer invokes a person, place, or animal. These days yoik performance also tends to serve as a marker of pride in Sámi identity.) The lyrical furthermore lives as an audio "map" in descriptive, poetic place-names. When everything is oral and is maintained by collective memory, all knowledge is living knowledge. The oral tradition plays central roles in communication and in the bearing of values and worldview. Its scope is vast, and contains the entire cultural heritage—laws and rules, worldview, literature (or perhaps oralture?), and economic knowledge—including everything from local botany, biology, meteorology, and knowledge of snow and ice conditions, river currents and marine biology, to art and handicrafts—as well as geography, history, and metaphysics.

The Sámi understanding of freedom that permeates our own approach and methodology must also have confronted Hamsun with a sense of dizzying possibilities. There is not only the freedom of the incomprehensible, but also a particular understanding of nature that includes human nature. In addition to the freedom *in* nature, there is a freedom to live *off* nature as a baby nurses its mother, to live with it in all its capers and whims. Perhaps the closest that Hamsun came to giving image to this desire was in his oft-repeated vagabond figures, the free tramps modeled on the pattern of the so-called beggar Lapp. Of course it was possible to live off the earth as a farmer, something he often praised, but it seems that we can also detect a bit of jealousy in the narrative revelations of a shocking lifestyle made up of simply drifting about in the mountains, apparently living off nothing—without shame, disgrace, or even a particularly strong need to beg at the settlers' farmsteads. What are we to make of this knowledge and magic that allowed an entire people to siphon sustenance directly from the soil, straight out of the mountains? It was yet another wonder, a puzzle from the Nordlándda wilderness, a freedom *not* to be bound to the soil, a freedom to drift, a freedom that even found expression in language, in the pidgin Norwegian of non-native speakers, in its small peculiarities, a lack of respect for language that enlivens it, making it festive, vital, and lively. It isn't even grammatically correct; but it works, and works better!

AMERICA AND ANARCHY

Our purpose is not to speculate too much about "the riddle of Knut Hamsun." But we are intrigued by the power in the opposition between the modern, fragmented subject and its longing for "the natural" and the bygone as ideals. This tension runs throughout the author's work, its impossibility foregrounded in his continual focus on the unconscious life of the mind.

In his years as a young man in America, Hamsun's experience of harsh working conditions, and his near drowning in debt, illness, and misery, inspired sympathies with the labor organization and workers' rights movements. Later, of course, as he rose in social station, he came to despise the workers' movement and denied ever having been a supporter (Ferguson 1988, 115), but in these early years he tended to identify with anarchist critics who decried the exploitation of immigrants as cheap labor. He wrote that, in America, foreign workers had taken the place of slaves, an observation that was no doubt grounded in his own bodily experiences as an immigrant laborer in Chicago and on the Dalrymple farm in the Dakota Territory (cf. Hamsun 2003a, 2003b).

In his 1889 book *The Cultural Life of Modern America* (*Fra det moderne Amerikas åndsliv*), Hamsun examines freedom in the United States. His basic premise is that it doesn't actually exist. Instead, he sees freedom in America as an illusion, a capitalist PR stunt created to keep the stream of ready-to-work immigrants flowing.

Spiritual and cultural freedom are recurring themes in Hamsun's work, and, though he was probably unaware of it, his ideal of freedom as personal and in close contact with nature—as simple but responsible—had a lot in common with Sámi ideals, too. (If Lt. Glahn needs a ptarmigan, for instance, he shoots one, not two.) "Freedom with responsibility" would be a decent way to describe traditional Sámi society well into modern times, something that, in turn, points to a resonance between Sámi society and anarchist theory.

When it came to another inhumane aspect of American capitalism, though, the way that Indigenous people were displaced to make room for the money economy and those that served it, Hamsun was further removed from Sámi sensibilities. He regurgitated worn-out tropes, acknowledging the tragedy that development foisted on the "Noble Savage," but seeing that development as inevitable and even necessary. He recognized the right of the strongest to expropriate land and water, to plunder and commodify the landscape and incorporate it into the capitalist system. He did not take the attempts by the local Indigenous people to negotiate with the authorities seriously, though, fortunately, the passage of time has shown that the tribes of the upper Midwest were in fact able to achieve a fair amount in these negotiations, to the benefit of current generations.

During his first stay in America Hamsun showed a good deal of sympathy for the civilization he mistook for Shawnee—but which was most likely Ho-Chunk (Žagar 2001, 2009, 263). And, while the episodes he narrates may well have been fictionalized, the fact remains that there were people living in the same area as him whose society was quite different from that of the immigrants, just as the Sámi had existed in the shadow of the Norwegian settlers back home in Hábmer. And like the Sámi, the Native Americans were seen by the settlers through thick lenses of stereotypes and prejudice. Yet even through these lenses, Hamsun's description conveys a sense of loss:

> It is with a certain sorrow that one observes the gradual decline of the Indians. They once owned the richest land on the planet, and were lords over a third of the world's land. They lived in their wigwams (tents) or out on their wide hunting grounds, hunted, fished, warred and took prisoners, and pleased themselves—as the forest's free sons and daughters. (Authors' translation; for the original Norwegian, see Hamsun 1998, 115)

It is easy to read the lyrical idealism that emerges here, though one blending with racialist ideology, to be sure. Yet, despite the stereotypes and prejudices, Hamsun approaches a recognition here—that of the plight of the colonized. He sees the proud memory of freedom maintained by people who *know*, who do not forget. Having traveled halfway around the world, he discovers something real—that this has to do with the right to land and water, and with the power of the people of money to take, develop, invest, and liquidate.

With the exception of the bit about being a warrior people, Hamsun could have written the same thing about the Sámi—if only they hadn't been so close to home. As a foreigner in America, he is better able to identify the power relations, the displacement of Indigenous peoples in favor of development and the people of money. The author's later treatment of the Wounded Knee Massacre, as expressed in his essay "*Red, Black, White*" ("Røde, sorte, hvite," 1891), is similar, and rather nuanced and progressive, given the dominant racist ideologies of the day. His language shows a distance and sobriety that appear carefully balanced, though he still adopts a passive relationship to what he continues to view as the inevitability of development. Despite his concern for the tragedy, though, he participates in the discourse of colonialism, maintaining and supporting the myths of the Wild Native and the Noble Savage.

Some years later, having returned to Norway, Hamsun contributed to the colonial discourse of nation-building there—the building of the Norwegian nation on the lands of the Sámi. Just as in America, the nation is built on land stolen from others, on the loss of others, and at their expense. Responsible, socially aware citizens are able to ignore the life-lie they live because the racialized order of knowledge obscures it. Perhaps it was to hold this recognition at bay that Hamsun turned to such reactionary extremes, drifting through life in search of something authentic, pounding the table with harsh opinions to keep this threat in check, to keep everything in place. Nowhere can this be seen as well as in *Growth of the Soil*, where Social Darwinism and colonial ideology cast long shadows, explaining "progress" and its stages of development as inevitable. At the same time, though, he's got his brakes on the whole way, trying to prevent just this inevitable progress and development. A stalemate is sought in an impossible paradox, in the hopes that it can hold everything in a safe embrace, where no escape is desired, where the bars are so beautiful and green that one completely forgets the longing for freedom.

COMING FULL CIRCLE

As we walked the mountain path between Strinda and Makkvatnet, in the footsteps of Hamsun's Sámi housekeeper Anna Pedersen ("Anna i Makkvatnet"), listening to the wind and the ptarmigan, and encountering reindeer from the Kalstad herd, a picture began to form of the triangular network of relationships that had inscribed this place a century earlier, when Hamsun sat in his room at Kråkmo, on a thriving backwoods farm surrounded by granite mountains, woodland bogs, and pine forests, visited regularly by Anna, a proud, self-sufficient Sámi woman who lived on her own on the other side of the mountain, and who, according to local tradition, was the only one who could put the Norwegian author in his place. The third point of the triangle, Strinda, was the base of the Kalstad reindeer herders who wandered past the farm, stopping for coffee and gossip on their way to see to their animals. This was the place that Hamsun had fictionalized as Sellanrå, the proud settler farm. But even though the Sámi had been banished from the area by the middle of the novel, in the place itself they remain to this day.

And perhaps even in the novel. After all, we found a number of similarities between Isak's wife Inger and the housekeeping Anna, with whom Hamsun had enjoyed a special relationship (according to local oral tradition, a very special relationship). While he went to lengths to purge Inger of her Sámi identity in the novel, Hamsun wasn't quite able to remove her komager (Sámi footwear) or her almond-shaped eyes. How did this believer in Norwegian racial superiority reconcile himself to his admiration for the strong Sámi woman from over the mountain? Did he find an outlet for his negative feelings by embodying them in Inger's kinswoman Oline (see also Jernsletten 2004, 2006)? Or might we turn to another of his characters, Åse of *The Road Leads On* (*Men livet lever*, 1933), to explore better the author's conflicted feelings about Anna?

As we sat in the parlor with Jorunn Kråkmo, whose late husband had been a boy during Hamsun's stays at the family farm, the past and present came together. Jorunn remembered Trine's father and uncles, who had herded in the area during her younger days, and the two of them exchanged stories of the local landscape and its people. She seemed to remember Uncle Nils especially well, her face glowing, her voice softening, and her eyes traveling back in time. As the light summer night wore on, Jorunn also told us stories of Hamsun and his irascible nature. Discussion gradually wound around to skirt on issues of Norwegian–Sámi relations, almost always a difficult topic in Sápmi. But in this case, it wasn't. Hamsun may have been conflicted on the topic, but the current owner of Kråkmo was not.

For Indigenous methodologies, the research process itself is often more important than the published results (Smith 1999, 128). In our various conversations in peoples' homes, in our meetings with colleagues at the Hamsun and Árran centers, and in our meeting with both Sámi and non-Sámi community members at Árran a year later, we were building and strengthening relationships. In these conversations with people, and in our conversations with the non-human residents of the place, we were engaging in what Shawn Wilson (Opaskwayak Cree) has described as research as ceremony. In his words, "the purpose of any ceremony is to build stronger relationships or bridge the distance between aspects of our cosmos and ourselves. The research that we do as Indigenous people is a ceremony that allows us a raised level of consciousness and insight into our world" (2008, 11).

In sharing our process here, we are broadening that ceremony further, to include you readers. Our next steps will be to publish a more extensive account of our research, including the methodological manifesto that we developed as we worked, in Norwegian and Lule Sámi, the local languages of the place with which we are engaging. It is of central importance that we return our knowledge to the community, and share with them our own perceptions of how Knut Hamsun, the sometimes racist giant of Norwegian national literature, is a product of the Sámi presence here, and now also another figure in local Sámi oral tradition.

WORK CITED

Acoose, Janice. 1995. *Iskwewak-Kah' Ki Yaw Ni Wahkomakanak. Neither Indian Princesses Nor Easy Squaws.* Toronto: Women's Press.

Ahluwalia, Pal. 2013. Derrida. In *Global literary theory: An anthology*, ed. Richard J. Lane, 81–93. New York: Routledge.

Anderson, Robert. 2010. The 'Idea of a University' Today. Last modified March 1, 2010. http://www.historyandpolicy.org/policy-papers/papers/the-idea-of-a-university-today. Accessed 23 Aug 2014.

Andreassen, Lars Magne. 2011. Skjult same i Narvik. In *Samisk skolehistorie 5: Artikler og minner fra skolelivet i Sápmi*, ed. Svein Lund, 390–401. Kárášjohka: Davvi Girji.

Armstrong, Jeannett. 2012. Sharing one skin. In *Asserting native resilience: Pacific Rim indigenous nations face the climate crisis*, ed. Zoltán Grossman, and Alan Parker, 37–40. Corvallis: Oregon State University Press.

Bergman, Ingela. 2008. Remembering landscapes: Sámi history beyond written records. In *L'Image du Sápmi*, ed. Kajsa Andersson, 14–24. Örebro: Örebro University.

Derrida, Jacques. 2002. Hospitality. In *Acts of religion*, ed. Gil Anidjar, 356–420. New York: Routledge.

———. 2005. 'There is no one narcissism' (autobiographies). In *Points ... Interviews, 1974–1994*, ed. Jacques Derrida and Elizabeth Weber, 196–215. Stanford: Stanford University Press.

Ferguson, Robert. 1988. *Gåten Knut Hamsun*. Oslo: Dreyers.

Fermantez, Kali. 2013. Rocking the boat: Indigenous geography at home in Hawai'i. In *A deeper sense of place: Stories and journeys of collaboration in indigenous research*, ed. Jay T. Johnson, and Soren C. Larsen, 103–124. Corvallis: Oregon State University Press.

Gaski, Harald. 1993. *Med ord skal tyvene fordrives: Om samenes episke poetiske dikt-ning.* Kárášjohka: Davvi girji.

———. 2015. Journeying with the son of the sun: South Sámi Yoik and literature in a Pan-Sámi perspective. In *Visions of Sápmi*, ed. Anna Lydia Svalastog, and Gunlög Fur, 149–170. Røros: Arthub.

Hamsun, Knut. 1992. *Markens grøde.* In *Samlede Verker, Bind 7*, 145–397. Oslo: Gyldendal Norsk Forlag.

———. 1998. Røde, Sorte og Hvide. In *Hamsuns polemiske skrifter*, ed. Gunvald Hermundstad, 115–116. Oslo: Gyldendal.

———. 2003a. On the Prairie. In *Knut Hamsun remembers America: Essays and stories 1885–1949*, edited and trans. Richard Nelson Current, 72–79. Columbia: University of Missouri Press.

———. 2003b. Vagabond days. In *Knut Hamsun remembers America: Essays and stories 1885–1949*, edited and trans. Richard Nelson Current, 91–121. Columbia: University of Missouri Press.

———. 2007. *Growth of the soil.* Trans. Sverre Lyngstad. New York: Penguin Books.

Ingold, Tim. 2000. *The perception of the environment: Essays on livelihood, dwelling, and skill.* New York: Routledge.

Ingold, Tim, and Jo Lee Vergunst. 2008. Introduction. In *Ways of walking: eth-nography and practice on foot*, ed. Tim Ingold, and Jo Lee Vergunst, 1–20. Burlington: Ashgate.

Jernsletten, Kristin. 2004. The Sámi in *growth of the soil*: Depictions, desire, Denial, *Nordlit* 15 (Special Issue on Northern Minorities), 73–89.

———. 2006. Det samiske i *Markens grøde*: Erfaringer formidlet og fornektet i teksten. In *Hamsun i Tromsø IV: Tid og rom i Hamsuns prosa II*, ed. Even Arntzen, and Henning H. Wærp, 117–138. Hamarøy: Hamsun-selskapet.

———. 2012. The hidden children of Eve: Sámi poetics: Guovtti ilmmi gaskkas. PhD dissertion. University of Tromsø.

Johnson, Jay T. 2013. Kaitiakitanga: Telling the stories of environmental guard-ianship. In *A deeper sense of place: Stories and journeys of collaboration in indig-enous research*, eds. Jay T. Johnson and Soren C. Larsen, ccf–138. Corvallis: Oregon State University Press.

Johnson, Jay T., and Soren C. Larsen. 2013. Introduction: A deeper sense of place. In *A deeper sense of place: Stories and journeys of collaboration in indige-nous research*, ed. Jay T. Johnson, and Soren C. Larsen, 7–18. Corvallis: Oregon State University Press.

Karlsen, Inga. 2011. En del av min oppvekst som same, *Bårjås* 2011: 24–32.

Kuokkanen, Rauna. 2007. *Reshaping the university: Responsibility, indigenous epis-temes, and the logic of the gift.* Vancouver: University of British Columbia Press.

Myrvoll, Marit. 2010. 'Bare gudsordet duger': Om kontinuitet og brudd i samisk virkelighetsforståelse. PhD dissertion. University of Tromsø.

saamicouncil. 2012. What local people? *Saami Resources.* Last modified January 3, 2012, http://saamiresources.org/2012/02/03/what-local-people/

Smith, Linda Tuhiwai. 1999. *Decolonizing methodologies: Research and indigenous peoples.* London: Zed Books.

Storfjell, Troy. 2003. Samene i *Markens grøde*—kartlegging av en (umulig) idyll. In *Hamsun i Tromsø III: Rapport fra den 3. Internasjonale Hamsun-konferanse, 2003; Tid og rom i Hamsuns prosa,* ed. Even Arntzen, and Henning H. Wærp, 95–112. Hamarøy: Hamsun-Selskapet.

———. 2011a. A nexus of contradictions: Towards an ethical reading of *Markens grøde.* In *Hamsun i Tromsø V: Rapport fra den 5. Internasjonale Hamsun-konferanse,* ed. Even Arntzen, Nils M. Knutsen, and Henning Howlid Wærp, 243–253. Hamarøy: Hamsun-Selskapet.

———. 2011b. Worlding and echoes of America in *Markens grøde* (*Growth of the soil*). In *Knut Hamsun: Transgression and worlding,* ed. Ståle Dingstad, Ylva Frøjd, Elisabeth Oxfeldt, and Ellen Rees, 189–203. Trondheim: Tapir Academic Press.

———. 2016. Dancing with the Stállu of diversity: A Sámi perspective. In *New dimensions of diversity in Nordic culture and society,* ed. Jenny Björklund, and Ursula Lindqvist, 112–128. Newcastle Upon Tyne: Cambridge Scholars Publishing.

Suchet-Pearson, Sadie, Kate Lloyd, Laklak Burarrwanga, and Paul Hodge. 2013. Footprints across the beach: Beyond researcher-centered methodologies. In *A deeper sense of place: Stories and journeys of collaboration in indigenous research,* ed. Jay T. Johnson, and Soren C. Larsen, 217–231. Corvallis: Oregon State University Press.

Tuorda, Tor Lundberg. 2014. What local people? YouTube. Last modified January 28, 2014. http://saamiresources.org/2012/02/03/what-local-people/

Wilson, Shawn. 2008. *Research is ceremony: Indigenous research methods.* Winnepeg: Fernwood.

Žagar, Monika. 2001. Imagining the red-skinned other: Hamsun's article 'Fra en Indanerleir' (1885). *Edda* 88(4): 385–395.

———. 2009. *Knut Hamsun: The dark side of literary brilliance.* Seattle: University of Washington Press.

CHAPTER 7

The Polar Hero's Progress: Fridtjof Nansen, Spirituality, and Environmental History

Mark Safstrom

In 1897, the Norwegian imagination received a real-life hero that seemed worthy of the sagas. In that year Fridtjof Nansen (1861–1930) published the account *Fram over polhavet* (*Farthest North*), which recounted the unbelievable feat of daring and athleticism by the crew of the *Fram* on its expedition in the Arctic between 1893 and 1896. In his narrative, the scientific findings of the expedition paled in comparison to its human existential dimensions. Realizing this, Nansen introduced his account by situating it against the backdrop of humankind's centuries-old search for philosophical and spiritual meaning (Nansen 1897a, 2); in the Norwegian preface, he even warns his readers of the subjective nature of his reflections. References to his own spiritual development are scattered throughout the text, and are prompted by his struggle with and submission to the natural environment. The conclusion of my study of his account is that this narrative often confounds assumptions about Victorian polar explorers, namely that they represented the triumph of rationality, athleticism, and male agency over the spiritual and "superstitious" medieval view of wilderness spaces. Even a recent anthology of Arctic exploration perpetuates these assumptions to some degree when such claims are made as that

M. Safstrom (✉)
Department of Germanic Languages and Literatures, University of Illinois at Urbana-Champaign, 707 S. Mathews Ave Rm 2090, Urbana, IL 61801, USA

© The Author(s) 2017
L.-A. Körber et al. (eds.), *Arctic Environmental Modernities*,
DOI 10.1007/978-3-319-39116-8_7

107

"the old Christian perspective on the wilderness was unequivocally negative" and that the nineteenth-century explorer/scientist represented "a
completely new perspective" (Wråkberg 2004, 23). Nansen, too, reflects
some of these modern assumptions, but overwhelmingly revives and
repurposes earlier traditions, as well. Most significantly in this regard,
Nansen situates his discovery in the context of religiosity in a traditional
European perspective, which was often coded as feminine and passive
(e.g., the church as the bride of Christ). By thus "feminizing" his account
with religious imagery drawn from medieval asceticism, Nansen assumes
a split personality of inner turmoil between conflicting ideals of heroism
and manhood.

The particular focus of this chapter is to present some ecological and
spiritual implications of how Nansen narrates his experiences of temptation in the Arctic. His narrative will be evaluated in light of Western
traditions of wilderness asceticism and compared with similar themes in
ecological philosophy, namely the "deep ecology" (or "ecosophy T") of
Arne Naess (1912–2009). Elements of Nansen's imaginative approach to
defining the wilderness are traceable in the writings of Naess, particularly
in such binary tensions as the spiritual versus scientific, feminine versus
masculine, and romantic versus rational. It is the hope that this investigation will contribute to the growth of a counter-narrative that identifies and
problematizes the persistence of the eco-spiritual traditions of wilderness
asceticism as they continue to inform and complicate the study of Arctic
cultures and literature.

Using the vocabulary of temptation as the common thread, I will seek
to answer the following questions. To what extent can *Farthest North*
facilitate an allegorical reading? If so, how does the account function as
a temptation allegory (narrative techniques, symbols) and what messages
does it convey (ideological, spiritual)? How does Nansen's account compare to the medieval allegorical traditions? How does his presentation of
the environment resemble that of deep ecology? And what relationship
(if any) can be made between the "progress" of the hero (Nansen) on
his journey through the wilderness and the potential for civilization (his
readers) to make progress toward the redemption of humanity from its
urban, industrial environment? I will suggest that Nansen's account does
function as allegory, offering readers a view of the Arctic as a stage for individual agency and accomplishment, as well as inspiration for corporate,
civilizational progress. The resulting picture of the Arctic is one in which

the reader-as-explorer has occasion to identify with Nansen as the polar hero, as well as identify with the environment that is tempting that hero.

WILDERNESS ASCETICISM AND DEEP ECOLOGY

Nansen's otherwise anthropocentric view of nature is complicated by the fact that his ascetic tendencies imply a limited ecocentric surrender of agency to nature, and thus inhabit a traditionally feminine/passive/religious posture. As his ecological view is so heavily defined through spiritual vocabulary, there is just cause to define his view as eco-spiritual, particularly to the degree that he is inspired by wilderness asceticism. To bridge the conceptual gap between the medieval monastic approach to the wilderness, and that of Nansen, the ecosophy of Arne Naess can be informative. Though they are distinct fields of human inquiry, wilderness asceticism and Naess's ecosophy have one significant thing in common: both promote a view of wilderness as being a space/entity that is ethically equal to or greater than human civilization. For the third century Egyptian ascetics, the wilderness offered a place of sanctification (a process of being made holy/perfect) precisely because it allowed for a renunciation of excess (wealth), but also because it involved spiritually uplifting opportunities to combat temptation (spiritual struggles against boredom, inactivity, and sexual temptation; and physical struggles against starvation and threatening beasts). For these spiritual athletes, "following Christ" meant renouncing the urban (or agrarian) environment and entering the wilderness. While the ascetic sought out redemption through a lifestyle of renunciation, there was also a corporate aspect. The ascetic could also serve as a prophetic symbol that forced the rest of society to confront their cultural errors and reform their ways, sometimes prompting monastic communities. Temptation experiences that resulted from immersion in and surrender to the wilderness were an occasion for criticism of societal materialism and excessive consumption (Louth 1997).

Deep ecology demonstrates a similar impulse, for different reasons. As Naess explains, the movement strives to decrease consumption and population growth in such a way that will simultaneously increase the quality of life. Implicit is an ideological regression or return to certain attitudes regarding consumption that can be seen as pre-modern (strategies for using less, making less). The result of this paradigm shift is to restore the equilibrium between humans and their environment. As Naess explains, Western society since World War II has demonstrated a systemic disrespect for "backward" cultures and worldviews, be they pre-industrial or religious. This bias particularly extends to subjective worldviews that personify natural phenomena or explain these phenomena with a spiritual vocabulary. The consequence of industrial society's preference for impersonal, objective definitions of the natural world is that this facilitates its overexploitation and destruction. The goal of deep ecology is to "change the picture" that industrial societies have of themselves (and that pre-industrial societies have of industrial ones). Furthermore this is a progressive regression that marks a return to sustainability, "not to old forms of society" (Naess 1995, 130–146).

Noteworthy is that Naess occasionally also uses the vocabulary of temptation to describe this struggle. He notes that "it is tempting to see 'us'—members of the rich industrial countries—as 'moderns,' more or less disregarding nine-tenths of humanity. They also live today, they belong to the contemporary scene, but are considered relics of the past" (Naess 1995, 143). Deep ecology is championed as having the potential to redefine modernity, if it conceives itself as a resistance against the allure of harmful ideologies and practices. The movement's goal is to progress toward a new society that is defined as "post-industrial," not retreat into a pre-modern existence.

The purpose here is to focus on how Nansen imagines his relationship to the other-than-human elements of the Arctic, and how this narrative resembles (and presages) that of Naess. Each demonstrates the idea that a re-identification with nature is necessary as an antidote to harmful, "modern" conceptions of selfhood. In the process of re-identifying with nature, Naess attempts to persuade his readers to see the benefit of the new "total-field image" and resist the temptation to accept the status quo. Nansen's temptations are presented as a process of self-realization, but one that has implications for societal views of humanity's relationship with the environment. Critics of both authors' notions of selfhood have pointed out that the identification of the self with nature has created a blindness to their gendered and national positions of privilege. Understanding the complex allegorical function of these texts can do much to alleviate that blindness.

FARTHEST NORTH AS TEMPTATION ALLEGORY

Apart from the primary reading of *Farthest North* as an expedition narrative, Nansen's account can also be seen as having a coherent secondary reading. Two things speak in favor of this. The first is that embedded in the narrative are references to other allegorical texts. When Nansen juxtaposes himself with Goethe's and Ibsen's heroes (e.g., Faust and Brand), he appears as a similar allegorical hero, the everyman wrestling with temptation in order to unlock his full potential. There is also a clear connection between Nansen's view of his role as a national hero and the "great man" theory of history as popularized by Carlyle. The second is that Nansen's vocabulary of temptation is congruent with the ways in which spiritual allegorists from Bunyan to Kierkegaard have explained such experiences. Contrary to what one might expect of a polar explorer, Nansen *seldom* describes the Arctic region and the North Pole as objects of temptation. Instead, the vocabulary of temptation he uses indicates that the strongest pull is almost always directed toward "home" (or a domestic, feminine self), or toward different potential versions of himself (the modern vs primitive self, etc.). In my close reading of both the Norwegian text and the English translation, I have catalogued the frequency of temptation vocabulary, including variants like longing, yearning, desire, allure, attraction, and seduction. Nearly one hundred times, Nansen makes use of such vocabulary, and the lion's share can be seen as deliberate choices (rather than colloquialisms).[1] Nansen has poetically recast his journey as an imaginative series of existential trials.

Identifying allegory depends on the degree to which Nansen's vocabulary can be seen as "coded speech," which assumes an awareness of multiple potential meanings on the part of reader and author. For readers of the period, there was a high level of familiarity with the allegorical traditions. At the time *Farthest North* was published, this tradition was experiencing a renaissance in Scandinavia, notably in the flurry of new translations of John Bunyan's *The Pilgrim's Progress*, originally published in 1678.[2] One explanation for this popularity is that readers resonated with this kind of allegorical protagonist, precisely at a time when social movements and religious revivals were emphasizing individual agency. Subjective soul-searching was a natural outcome of movements that stressed the importance of individual decision-making in the process of reforming society. Differentiations between the self and the community were current in literature and philosophy, perhaps most radically articulated by Søren Kierkegaard. Relevant here is the role that isolation and temptation played in Kierkegaard's presentation of the soul's progress toward maturity. "Temptation" (*anfægtelse*) was articulated as a multi-directional temptation, a moment of despair on an interior journey, as the individual is tempted toward the destination, away from the destination, and even sometimes repelled by it. This is a severe and recurring crisis, but is essential for the individual to be able to identify subjectively with the truth of lived experience.

A natural connection between Nansen and this tradition of temptation is through Ibsen's play *Brand* (1865), which is partially inspired by Kierkegaard's Abraham in *Fear and Trembling* (*Frygt og Bæven*, 1843). There was a copy of *Brand* in the library of the *Fram*, and the library catalog indicates that Nansen checked out the book during the expedition (*Katalog over Frams bibliotek*). With this in mind, when one examines Nansen's language, there is a resonance with Kierkegaard's concept of temptation. This is apparent as Nansen expresses his restlessness and rationalizes his desire to leave the ship and make an attempt for the pole (Nansen 1897a, 227). An overarching theme is that temptation and longing are a productive experience: "But longing—Oh, there are worse things than that! All that is good and beautiful may flourish in its shelter. Everything would be over if we cease to long" (Nansen 1897a, 217).

Nansen's admiration for Thomas Carlyle's ideas concerning heroism is clear in his explicit reference to *On Heroes, Hero-worship, and the Heroic in History* (Nansen 1897a, 2: 46). Published in 1841, these essays by Carlyle highlight a series of historical examples making the case that humanity needs heroes to pave the way for the rest of civilization to follow and make progress. Carlyle's praise of Norse mythology, including his allegorical reading of Thor's "expedition" to Jotunheim, and his presentation of "Mohamet's" experiences and education in the desert seem to have resonated with Nansen most, as Norse mythology and references to Islam abound in *Farthest North* (Carlyle 1893, 41, 58). Nansen's ideas about the hero's relationship to nature also suggest inspiration from Carlyle. According to Carlyle, "man first puts himself in relation with Nature and her Powers, wonders and worships over those." As the primitive hero ("the First Norse 'man of genius'") realizes the moral implications of his struggle with nature, he is awakened and develops into "the Thinker, the spiritual Hero" (Carlyle 1893, 24). Thus inspired by a wonder of nature, the hero begins as "a minority of one," and must struggle to establish his ideas among the rest of society: "In this great duel, Nature herself is umpire, and can do no wrong: the thing which is deepest-rooted in Nature, what we call truest, that thing and not the other will be found growing at last" (69). As part of this struggle, the heart of the hero will face "allurements" (difficulty, abnegation, martyrdom, and death) that would pull him off course and prevent his progress and that of the race (79).

Nansen's particular reading of Carlyle is congruent with Naess's deep ecology on a couple of points. Naess makes frequent use of the similes "life is like traveling" or an "expedition in the mountains," and expeditions into nature and the sense of wonder about nature are key ingredients in the individual's progress toward emotional maturity (Naess 2002, 1–2, 35, 94). Identification with nature involves imaginatively placing oneself in relation to nature, which allows nature to "help" the individual to view the "total field image" of the interconnectedness of all beings (20). In contrast with Carlyle's violent image of nature, Naess's mountain hut Tvergastein presents a benign image of nature as a retreat, but a retreat that is necessary for progress. Naess finds inspiration to reclaim certain Enlightenment principles, namely Spinoza's "belief in the possibility of the individual's making progress" (75), but cautions that these Enlightenment ideas need to be tempered with "deeper, value-oriented premises" (63). These premises include the (re)introduction of feeling and emotion into the discussion of scientific research and environmental sustainability. The preference for cold rationality within science has created a bias against emotion, and Naess makes the case that emotional maturity is necessary for humanity to make progress toward sustainability and individual fulfillment. Wrestling with temptation factors into this maturity, as social patterns and internalized community rules must progress from being external obligations to being internalized values (121). The individual will experience temptations to evade responsibilities to the network of all living beings, but can overcome these temptations as he or she matures. One primary difference with Nansen is that Naess has little place for lone heroes. The maturity of the individual self is cultivated through temptation experiences, which blossom as deep ecology becomes a collective movement.

IDENTIFYING AN ARCTIC ASCETICISM

According to the common assumption regarding the medieval view of the wilderness, nature represented the great adversary to humankind, to be conquered with the ax and plough. The woods, groves, high places, and glaciers were inhabited by evil spirits or associated with abandoned pagan practices (see Wilson 2003 for contrasting perspectives). The Victorian explorer/scientist, by contrast, supposedly championed a new perspective, which elevated the wilderness as a place for athletic accomplishment and spiritual renewal, as well as scientific advancement and resource extraction. As mentioned already, a recent Gyldendal anthology of Norwegian polar exploration begins with such a prelude, asserting that the polar explorers demonstrated a rejection of the medieval mindset (Wråkberg 2004, 23). This assumption has some validity, yet warrants an important qualification. This same Romantic tendency that exalted the wilderness did so by recycling and repurposing, rather than rejecting, earlier ascetic traditions. Nansen is a prime example of this. Christian asceticism utilized biblically based interpretations of the wilderness as a space that, despite being hostile, was formative for positive spirituality. For those who submitted themselves to it as a discipline, the wilderness could become a space for introspection, clarity, purification, and preparation for action. This spiritual discipline succeeds when it facilitates a process of stripping away the familiar and the distracting. The Arctic is ideal, since its extreme conditions strip away the basics of civilization, even warmth and sustenance. Ascetic vocabulary starkly delineates spaces, coding them as civilization/ wilderness, unclean/clean, profane/holy. Entering such spaces involves ceremonies of purification, or at least a state of mind that appreciates the transition. Nansen demonstrates familiarity with this practice. Before leaving the last outpost of civilization, he describes the crew's final opportunity for bathing in terms of a ritual of purification, and wonders if Mahomet had "a bath house" like this in his paradise: "[The crew's bodies undergo] one last civilized feast of purification, before entering on a life of savagery" (Nansen 1897a, 69).

However, unlike the desert fathers, solitude for Nansen is scarce. As such, it has to be manufactured and imagined. With books as his ultimate retreat, Nansen ironically complains of being "alone," with only nature and books to keep him company (Nansen 1897a, 309). The reader may wonder what happened to the other dozen crew members. Whereas the Arctic was originally a space for retreat, now the library serves as a retreat from the Arctic and other people; "a little oasis ... in this vast ice desert" (152). The act of reading was a compound temptation for Nansen, one which had evolved since childhood. As a boy, reading British expedition accounts had filled him with admiration: "all my boyish fancies were strangely thrilled with longing for the scenery and the scenes which were displayed before me" (Nansen 1897a 2: 14). Now actually in the Arctic, he grapples with the temptation to retreat into the life of the mind, instead of going out into nature (341).

Since it can be viewed as a retreat from reality, asceticism has a negative reputation in reform movements. Naess is careful to explain that the simplicity he is prescribing is not to be confused with asceticism, pointing out how Ghandi was also opposed to that strategy (Naess 2002, 169). However, he admits that diverging from the dominant norms requires some variant of the ascetic lifestyle: "to live in an ecologically sustainable way in a society like ours does not perhaps demand that one be a hermit, but that one be a social deviant of an unusual kind" (129). The prioritization of time spent deep in nature, whether at Tvergastein or on an expedition in the Himalayas, seems to be a variant of asceticism, which allows the individual to mature into the "right kind" of deviant. In both cases, the disciplines recommended by these authors involve imaginary constructions of the wilderness that often do not acknowledge what is being omitted from the picture (such as other human beings, traces of civilization, impurities that contradict notions of environmental purity, or the privileged position of the observer).

COMPLEXITIES AND CONTRADICTIONS OF NANSEN'S TEMPTATIONS

What is noteworthy about Nansen's encounter with temptation in the Arctic is that the strongest pull is *not* toward the North Pole, but toward home. Nansen dismisses the importance of reaching the exact mathematical point of the pole, and uses the spiritually coded term "vanity" (one of the seven deadly sins) (Nansen 1897a, 224). The pole seemingly takes on a negative charge, effectively repelling Nansen back home:

> But, O Arctic night, thou art like a woman, a marvelously lovely woman. Thine are the noble, pure outlines of antique beauty, with its marble coldness. ... Oh, how tired I am of thy cold beauty! I long to return to life. Let me get home again, as conqueror or as beggar; what does that matter? But let me get home to begin life anew. (213)

Nansen's temptations serve to redirect his attractions. While the male athlete might be expected to embrace hardship and forsake the domestic and the comfortable (thus the feminine), Nansen often makes home the ultimate goal. The female body, which should attract, here repels and confuses common sexualized metaphors for the Arctic as conquest.

There are also instances where Nansen identifies with the non-human world as a means of understanding his own temptations. On multiple occasions, he reflects on the plight of the sled dogs, and consequently contemplates Darwinian theory. Included in the library of the *Fram*, and read by Nansen, was a copy of Charles Darwin's *On the Origin of Species* (1859). On numerous occasions, Nansen philosophizes on the experience of the dogs; the bitter cold, bear attacks, infighting, and the cruelty the dogs receive at the hands of their caretakers. The dogs are the ultimate tragedy of the expedition, as one by one they are killed to be fed to the remaining dogs. Here, Nansen allegorizes their plight as a reflection of human temptations. The open conflicts between the dogs themselves hint at the hidden tensions between the "civilized" crew members, and this struggle also mirrors that of the allegorical hero, as he resists the temptation toward savagery and competition on the one hand, and toward compassionate ethics on the other. The Norwegian edition included two illustrations, which depict the range of emotions of the dogs. In the first sketch, two dogs are seen blissfully napping, tenderly nuzzled against one another, titled *To venner* (*Two Friends*). This is then juxtaposed with another image, called *To fiender* (*Two Enemies*), in which two dogs are viciously at each others' throats, and two crew members attempt to separate them, pulling them by their tails and suspending the dogs mid-air, like a living tug-of-war rope (Nansen 1897b, 420–421).

The slim margin of subsistence demands that the explorers conform to a sustainable lifestyle, and this environmental constriction transforms Nansen into a new kind of hunter. Early in the account, Nansen finds himself romanticizing the lifestyle of the nomadic peoples in Russia. As he describes the allure of nomadic life, there is a moment of temptation to escape civilization, and live the free life of the state of nature: "[the nomad] has no goal to struggle towards, no anxieties to endure—he has merely to live! I well-nigh wished that I could live his peaceful life, with wife and child, on these boundless, open plains, unfettered, happy" (Nansen 1897a, 81). As Nansen grows accustomed to hunting Arctic game, there is an attempt to mimic the patient stalking and deadly strike of the wild animals and the restraint of the nomadic hunter. Whereas the European gentleman hunter might shoot at everything that moves while growing bored waiting for the "big game," Nansen rejoices as he is able to resist this temptation and fully enjoy the thrill of "the strike" as he scrambles over rocks and crawls through miry clay like a "wild beast" (108). Later, when Nansen and his companion Hjalmar Johansen are fully dependent on their ability to hunt, this triumph of the primitive, sustainable hunting practices appears as a development toward maturity. All of these conflicted and confused metaphors make it clear that Nansen's temptations and longings are equally complex.

Educational Aspirations of Pluralistic Eco-Spirituality

Early in the journey, Nansen comments that, also among indigenous peoples, religious schisms have "found their way" (Nansen 1897a, 85). Here he reveals an assumption that the northern periphery of civilization would be a pristine, schism-free, religious desert. However, on the way to and from the 86th parallel, he ironically carries the baggage of millennia of religious thought farther north than anyone before. His unbridled choice of spiritual subject matter and playful use of imagery becomes an exploration of pluralism, which approaches a critical, open-minded agnosticism. Since these religious images are already pregnant with cultural meaning, they lend themselves to being altered for new allegorical purposes. Nansen's baptism of the Arctic with religious imagery is a creative process in which he interprets the tensions of being located in this landscape and communicates these feelings to his readers in comprehensible ways with old vocabulary. This agnosticism demonstrates a move away from a dogmatic atheism, toward an open-minded inquisitive posture, and potentially a humbler, less anthropocentric, worldview.

The pluralistic imagery that Nansen uses to identify with the environment resembles the pluralism advocated by Naess, who asserts that diversity of religious views does not conflict with deep ecology. Rather, its supporters "coalesce in one movement in spite of their differences ... there is diversity, luckily, not consensus" (Naess 2002, 6). Conventional Western science has effectively developed a culture that has demanded that the imagination be "tamed," which Naess explains "reduces the creative power of humankind" (71): "for a long time a clearly expressed gratitude to God was a constant ingredient in strictly scientific literature. Now the style of periodical literature has become more or less completely arid" (68). In the interest of promoting the emotional maturity of the individual, he prescribes that "we sorely need to nurture our mythlike imagination," going so far as to admit that he allows himself to imagine Hallingskarvet Mountain as being "alive," finding inspiration in the traditions of the Himalayan peoples (111). As Naess makes his case for a "convergence of reason and feeling" in industrial society's re-identification with nature, a key component of this process is the creation of an educational environment in which students can feel a burning "fervor" for something (85, 147). Academic interest in biology as learned in a textbook must be replaced with actual lived encounters in nature, as a means of honing general interests into mature passions (an argument for courses in the field).

Such fervor is also necessary for deep ecology to be a movement, rather than an academic discipline. Naess's "formula for well-being" articulates that this fervor is refined by mental and physical suffering (179). Similarly, Nansen appears to have drawn inspiration from Carlyle's praise of Dante's epic poem *The Divine Comedy* (1320), which highlighted Dante's theme of becoming "perfect through suffering" (Carlyle 1893, 102). This may explain in part why Nansen's journey through the Arctic is conceptualized as a journey through his own individual purgatory. When he overtly frames his entry into the Arctic by describing it as the Great Ice Church (a reference to *Brand*), it is also possible to conceive of his journey as a loop of stations similar to the "stations of the cross" in Catholicism. Nansen's meditations on suffering become opportunities for character building, with lessons being learned at each station. Among other virtues, the Arctic becomes a "school of patience" (Nansen 1897a, 312), and he even names one camp in honor of his longing "Længselens Leir" (Nansen 1897b, 2:149). Thus, the prescribed practice of both authors is for the individual to imagine a passive submission to the suffering that is imposed by the dominant natural environment.

Common to many classical and medieval allegories is the concept of the "daemonic," an idea Nansen also invokes:

> What demon is it that weaves the threads of our lives, that makes us deceive ourselves, and ever sends us forth on paths we have not ourselves laid out, paths on which we have no desire to walk? Was it a mere feeling of duty that impelled me? Oh, no! I was simply a child yearning for a great adventure out in the unknown. (Nansen 1897a, 338)

Angus Fletcher explains the "daemonic" as an obsession with one idea, or a single-minded purpose, not necessarily negative (Fletcher 1964, 40–68). Nansen's trials allow him to discern the purpose of his journey, his life, and the meaning of his interaction with the environment. His temptations include various self-deceptions, fluctuating moods, and the wavering belief in his own agency or fate. Not content to wait for the ice to move them south, he thirsts for struggle and action. Though his microscope and research "tempt" him, he exclaims that he would gladly give them up for action (Nansen 1897a, 304):

> Longing, even when it is strong and sad, is not unhappiness. A man has truly no right to be anything but happy when fate permits him to follow up his

ideals, exempting him from the wearing strain of every-day cares, that he may with clearer vision strive towards a lofty goal. (223)

Like Kierkegaard's complaints about the decadence of the age, and the lack of passion "to will one thing,"[3] Nansen similarly concludes that "everything seems to become more and more indifferent. One longs only for one single thing" (Nansen 1897a, 2:187). Nansen's cure for his own divided attentions is to isolate himself so deep in the Arctic that there is nothing else that matters except returning home. The "south" becomes the new goal, evident in the way in which he describes Franz Josef Land and Scandinavia with warm, Mediterranean imagery (Nansen 1897a, 327, 2:219, 2:222, 2:228). The world map has been reoriented and flipped, and the reasons for trying to reach the pole have changed. Now the "farthest north" point is valuable because it is the one spot that can help Nansen overcome his apathy and re-identify with his southern home. The Arctic forces the individual to become apathetic to everything that is not the one worthy goal: survival and return (324). This necessitates the elimination of opportunities for retreat (346).

Nansen's preface provides the readers with a disclaimer concerning the subjective nature of his narrative (lacking in the English edition). The readers are warned not to expect a scientific report, but rather a description of the experience of living in the Arctic (Nansen 1897b, preface).[4] Elsewhere, he notes that it has been in the act of writing that he has been able to understand his internal experiences: "the only thing that helps me is writing, trying to express myself on these pages, and then looking at myself as it were from the outside" (Nansen 1897a, 228). The act of writing down his scattered, ever-changing labyrinth of moods brings them into an exteriority in which he (and the readers) can interpret them. Without this journaling process, his disparate experiences might not have a discernible meaning, which he fears would be a "nightmare" (230). However, while he has attempted to look at himself "from the outside," he does not allow enough critical distance to draw conclusions about his tangled web of mixed metaphors, his complex gendered relationship to the natural environment, and his messy definition of selfhood. He remains the privileged gentleman hero, even though his own description of his experiences have inherently called this into question.

Toward a Critical Appreciation of Arctic Asceticism

Nansen's allegorical style seems to have left a noticeable impact on subsequent ecological philosophy. To be sure, plenty of differences exist between Nansen and Naess. Paramount among these is that Nansen attempts to manifest in his own person the solitary, heroic ideals of the medieval allegorical tradition, as refracted through Kierkegaard, Ibsen, Goethe, and Carlyle. Naess, on the other hand, champions a collective movement of thousands of little heroes, resisting temptations of the status quo to make progress toward a sustainable future. Similarities are evident in the naming strategies by which both attempt to identify the self with the environment. For both, experiences in nature become understandable to the modern industrial human by being re-enchanted into a fantasyland of imagery borrowed from the world's religions. Nansen's mixing of metaphors may risk being confusing, but at a basic level can be seen as an effort to achieve the kind of universality necessary for a diversity of readers to be able to find meaning in his artistic representation of the Arctic.[5]

For Naess, pluralism is a hallmark of intellectual humility and emotional maturity, but is also a practical strategy that can steer the ecological movement away from devolving into doctrinaire, counterproductive sectarianism (Naess 1989, 89, 91). Naess sees the Norwegian outdoor traditions (*friluftsliv*, heavily informed by Nansen) as bearing a "vital" role in developing a mature environmental paradigm (177–181). His openness to religious worldviews is in part due to their ability to create holistic characterizations ("gestalts") of natural phenomena and organisms, through a metaphoric process of naming that "bind the I and the not-I together in a whole" (60). Such multifaceted, mature, emotional connections to nature are what he finds lacking in the industrial paradigm. He explains that "if the developer could see the wholes, his ethics might change. There is no way of making him eager to save a forest as long as he retains his conception of it as merely a set of trees" (66). With this change of paradigm, humans would be able to see themselves not as "a thing in an environment, but a juncture in a relational system" (79). Belonging and kinship with the environment is facilitated through these practices of naming, whereas "the glorification of conventional 'scientific' thought leads to the ridicule of such creations" (61). In other words, the mature emotional ability to be filled with imaginative, holy wonder at an Arctic landscape is presented as being directly related to the imagination necessary to "change the picture" of modern society and facilitate progress toward a sustainable future.

Among the conclusions that can be drawn is that there is a high degree of continuity between the traditions of wilderness asceticism and both of these modern authors. While Nansen has been subject to more critique as the privileged, white, male explorer-hero à la Carlyle, these conceptions of heroism are easily revived, and they persist in Naess's writings, as well. Both advocate an imaginative relationship with the natural world, but at the same time do not substantially problematize the national and gendered implications of these imaginaries. As neither author distances himself critically from the conflation of selfhood with nature and with other selves, their readers may have similar difficulty in doing so. The "ecosophy T" of Naess has been embraced by those who find this philosophy useful in their attempt to correct the negative environmental impacts they see stemming from industrial society's excesses. However, the very usefulness of this philosophy may have to do with the fact that it builds on traditional binaries, without sufficiently examining their complexities. Thus the average resident of Oslo or any other industrial city can "nurture their mythlike imagination" on a weekend hike or ski trip, without needing to challenge the world order of their weekly existence or their privileged position. Without critical distance, dabbling in this kind of unreflective quasi-spirituality becomes as much entertainment as hiking. This imaginative play-acting may also obscure the privileged positions of secular ideologies, when traditional religious ideologies are seen as primitive, quaint, and anti-modern. Furthermore, there also remains a sense that, even though the narratives of Naess and Nansen are coded as agnostic and secular, the eco-spiritual roots of these discourses are profound, productive, and problematic. If Arctic-related scholarship, philosophy, and literature are to move in the direction of a more critical understanding of spiritual vocabulary, this will necessitate a fuller awareness of the persistence and relevance of spiritually formulated conceptions of the natural environment in all their many, complex forms.

NOTES

1. This study is based on 96 examples from the English text and 81 from the Norwegian. It may be that the English translation accentuated the vocabulary of temptation, though this study is not exhaus-

tive. In English the single most frequent word was "longing," and in Norwegian "*lengsel*," as well as variations on those words.

2. Published in Danish and Norwegian as *En Pilegrims Vandring*, and in Swedish as *Kristens resa*. Historian Gunnar Westin noted in 1924 that there had been at least 20 Swedish translations (not simply editions).

3. Kierkegaard expanded on this notion of "willing one thing" in "Upbuilding Discourses in Various Spirits" in (2000b) and "On My Work as an Author" (2000a).

4. "*Jeg har bare to ting at si: For det første at denne bog er blit for personlig farvet til at være en reiseberetning i almindelig forstand. Men jeg har det håb at fortællingen dog vil skinne gjennem det subjektive, og at der fra stemningernes skiften vil dæmre frem et billede af naturen og livet i den store isensomhed*" (Nansen 1897b, preface).

5. Angus Fletcher explains that the universality of an allegory depends on its "catholicity," in other words, its ability to be understood by all (drawing on Tolstoy's definition of "catholicity" of art) (Fletcher 1964, 327).

Work Cited

Bloch-Hoell, Nils E. 1984. *Fridtjof Nansen og religionen – Kirken og Fridtjof Nansen*. Oslo: Det Norske Videnskaps-Akademi.

Carlyle, Thomas. 1893. *On heroes, hero-worship, and the heroic in history*. New York: Frederick A. Stokes.

Diehm, Christian. 2002. Arne Naess, Val Plumwood, and deep ecological subjectivity. *Ethics and the Environment*. 7(1): 24–38.

———. 2007. Identification with nature: What it is and why it matters. *Ethics and the Environment*. 12(2): 1–22.

Fletcher, Angus. 1964. *Allegory: The theory of a symbolic mode*. Ithaca: Cornell University Press.

Huntford, Roland. 1998. *Nansen, the explorer as hero*. New York: Barnes & Noble.

Katalog over Frams bibliotek. Fram museum, Oslo. Katalog bibliotek Fram I – 119647A. Archival material, received as a PDF file to author.

Kierkegaard, Søren. 2000a. On my work as an author. In *The essential Kierkegaard*, ed. Howard V. Hong, and Edna H. Hong, 449–481. Princeton: Princeton University Press.

———. 2000b. Upbuilding discourses in various spirits. In *The essential Kierkegaard*, ed. Howard V. Hong, and Edna H. Hong, 269–276. Princeton: Princeton University Press.

Louth, Andrew. 1997. *The wilderness of God*. Nashville: Abingdon Press.

Naess, Arne. 1989. *Ecology, community and lifestyle: Outline of an ecosophy*. Trans. and revised by David Rothenberg. Cambridge: Cambridge University Press.

———. 1995. Industrial society, postmodernity and ecological sustainability. *Humboldt Journal of Social Relations* 21: 130–146.

———. 2002. *Life's philosophy: Reason & feeling in a deeper world*. Athens: University of Georgia Press.

Nansen, Fridtjof. 1897a. *Farthest north*. London: Macmillan & Co.

———. 1897b. *Fram over polhavet*. Kristiania: H. Aschehoug & Co.

Westin, Gunnar. 1924. Foreword in *Kristens resa*. Trans. G.S. Löwenhielm. Stockholm: B.M:s Bokförlags A.B.

Wilson, Eric. 2003. *The spiritual history of ice: Romanticism, science, and the imagination*. Gordonsville: Palgrave Macmillan.

Wråkberg, Urban. 2004. Polarområdenes gåter. In *Norsk polarhistorie. Vol. I, Ekpedisjonene*, ed. Einer-Arne Drivenes and Harald Dag Jølle. Oslo: Gyldendal.

CHAPTER 8

Heritage, Conservation, and the Geopolitics of Svalbard: Writing the History of Arctic Environments

Dag Avango and Peder Roberts

This is a chapter about the conceptualization, representation, and management of Arctic environments (natural and cultural) and how those processes inevitably always say more about people than about places. Our central argument is that imaginations of Arctic spaces were, are, and always will be snapshots of cultural, political, and economic geographies as much as physical geographies. The anxieties and ambitions of particular people at particular times are inscribed upon environments through description and location within narratives that in turn provide frames for practices. We therefore argue for the need to spend more time and effort analyzing how and why the natural and cultural landscapes of the Arctic are represented as they are—today as in the past—and what wider purposes those representations serve, while regarding any claims that things are natural or inevitable with suspicion.

D. Avango (✉) • P. Roberts
Division of History of Science, Technology and Environment, KTH Royal
Institute of Technology, Teknikringen 74D, 10044 Stockholm, Sweden

© The Author(s) 2017
L.-A. Körber et al. (eds.), *Arctic Environmental Modernities*,
DOI 10.1007/978-3-319-39116-8_8

While heartened that humanities scholars (like scholars of all stripes) have shown increasing interest in the Arctic of late, we want to stress the importance of not exceptionalizing the Arctic. Beau Riffenburgh (1993), Francis Spufford (1997), Urban Wråkberg (1999), and others have written about the nineteenth-century construction of the Arctic as a paradigmatic space of the sublime, in which the cultural vocabulary of European romanticism was inscribed upon the icy lands and seas of the north with little discrimination. Images of Indigenous peoples living outside the parameters of Western progress persisted well into the twentieth century, underpinning ideologies of colonial governance in Danish Greenland and elsewhere. In the present the Arctic is often represented as a place "open for business" in which the commercial environment takes center stage, often described as a consequence of climate change impacts in the region—notably the decreasing extent of the Arctic Ocean sea ice (Avango and Högselius 2013). These narrations say just as much about the people doing the telling as the places they describe.

We find value in the field of critical geopolitics, as pioneered by Gerard Toal (1996) and others, because it emphasizes how environments are always constructed within, rather than being external to, narratives of human activity. Knowing and representing a space can never be separated from controlling it. Even the technologies of representation employed by the natural sciences—ice core measurements, geological mapping, and so forth—never speak for themselves: they are always placed into context by people. Environments are always described within the context of narratives that give meaning to those spaces—and validate particular courses of action. Animals and ecosystems, as well as material remains of human activities, can be attributed qualities such as vulnerability that demand humans to speak for them. Discourses of environmental management and cultural heritage management are exercises in power that ought never to be regarded as neutral, benevolent acts, even when clothed in the language of ecology or presented as attempts to save a unique common heritage of humanity.

The empirical heart of this chapter is the Spitsbergen archipelago—today known as the Norwegian province of Svalbard. Spitsbergen has long been constructed as a resource base (for whaling, hunting and later mining), as a wilderness that Norway is responsible for protecting, as a space for international science, and much else besides. The physical geography of Spitsbergen has never dictated a uniform political, cultural, or economic response. Unlike so many other Arctic spaces, Spitsbergen lacks an indigenous population—although actors from Norway, Russia, and elsewhere have worked strenuously to manufacture national belonging (Avango 2005; Roberts and Paglia 2016). Our primary concern in this chapter is to explore how Arctic spaces are constructed by Europeans rather than by indigenous Arctic residents. While all understandings of environments are located within broader narratives, our lack of expertise in indigenous studies means we leave important questions for others to address.

NATURE, NATION, AND CONSERVATION

Debates over natural resource extraction on Spitsbergen relied upon a conception of the archipelago's physical geography as requiring incorporation within wider political and economic geographies. Entrepreneurs and scientists from Europe had envisioned it as a source of raw materials for Northern European economies even by the second half of the nineteenth century, but despite several attempts, nothing had come of these ventures. In the opening years of the 1900s, however, mining industrialists with substantial capital and experience were able to realize their visions. The Spitzbergen Coal and Trading Company, backed by British capital, founded the coal-mining settlement of Advent City in 1905. The following year the American mining entrepreneur John Munro Longyear (1850–1922) established Longyear City (today known as Longyearbyen), which to this day remains the most populous settlement on the archipelago. The success of Longyear's mining operation spurred companies from Russia, Sweden, the Netherlands, Norway, and Britain to establish coal mines on Spitsbergen during the decade that followed, envisioning that the mines would supply a growing energy market in the industrializing northern part of Europe. This emergence of the archipelago as a site of economic activity in turn spurred a sense that some form of legal order was required to facilitate effective operations.

The imposition of that order could also be an end in itself. The nineteenth century was an age of nationalism across Europe, including in Norway, which until 1905 was yoked to Sweden through a union with the Swedish crown. A generation before independence, Norwegian politicians were already concerned with making Spitsbergen a Norwegian territory (Berg 1995, 2004). The political maneuvering through which the archipelago became the Norwegian province of Svalbard was framed within a narrative of legitimate Norwegian authority. This narrative demanded knowledge of physical geography to naturalize Norwegian political power. Geological surveys, observations of fauna distribution, and the creation of national parks each contributed to this goal. So did the establishment of mines controlled by Norwegian companies, in several cases initiated by Norwegian geo-scientists (Avango 2005).

The most important figure in this process was Adolf Hoel (1879–1964), who first visited Spitsbergen in 1907. He had excelled in geology at the University of Kristiania under the mentorship of Waldemar Christopher Brøgger, whose stint as professor of geology in Stockholm only strengthened a deep commitment to viewing science as an expression of patriotic attachment to territory (Hestmark 2004). Hoel's interest in Spitsbergen coincided with Norway's independence from Sweden in 1905 and a growing feeling in political circles that the archipelago ought to be annexed (Berg 1995, 150–7), but that interest did not translate into funding, even for cartography—a prerequisite for effective administration (Drivenes 2004, 177). The quest to create demand for research on Spitsbergen—and to link that research to Norwegian sovereignty over the archipelago—became the twin themes in the geopolitical narrative that framed Hoel's career.

Hoel combined a commercial and a scientific gaze upon the environment of Spitsbergen. The coal deposits he discovered in 1907 led to another expedition the following year that combined geology and botany with a private claim to a site Hoel deemed particularly promising (Hoel 1966, 738). Like many other geologists on Spitsbergen at this time, Hoel took a personal interest in converting fieldwork into potential commercial results—even creating companies for the sole purposes of exploring the archipelago and occupying promising sites (Hoel 1966, 840–3), a task that could be understood as implicitly approved by the Norwegian government (Drivenes 2004, 196). The expertise of the geologist not only separated the valuable from the worthless in a financial sense (such as gypsum being mistaken for marble), but also in a moral sense, providing a rational foundation for exploitation (Hoel 1928, 1). This line of reasoning helped depict a space hitherto outside the sovereign authority of a state as requiring the imposition of legal and political order. As the interpreter of the environment and its resources, the geologist could reveal the natural order that in turn permitted the state to impose rational administration.

The need to impose order over nature reflected a simultaneous discourse demanding the imposition of order over people. Marco Armiero's work on mountains and nationalism in Italy has demonstrated that rugged Alpine environments were viewed as sources of superior citizens, who were assets to the nation's defence akin to the mountains from which they came (Armiero 2011). But unpopulated Spitsbergen was a crucible rather than a cradle. Legitimate authority derived not from the innate rights of its indigenous residents, but from the superior quality of its occupiers and administrators. In the history of Svalbard that Hoel wrote late in his life, he recalled how the (British) manager of one mining camp became "a virtual prisoner" of unruly, frequently drunken, mine workers in 1907, and blamed it on the company's representatives not being sufficiently careful in their choice of employees—many of whom were "well known to the police in northern Norway" (Hoel 1966, 564). Hoel claimed that disputes over who controlled what field sometimes led to threats of fatal violence, reinforcing his view that people needed order, and that Norway was the logical state to provide it (Hoel 1966, 652–3). Once Spitsbergen became Svalbard in 1925 the existence of a clear administrative framework created a stable environment for economic activity that in turn could legitimize Norway's sovereignty: Hoel's account of a dispute being settled through lawyers and the Norwegian Department of Commerce in 1927 was a far cry from the Wild West atmosphere of two decades prior (Hoel 1966, 845).

The advent of centralized authority also increased the importance of accurate mapping, as registering claims was a matter of direct administrative relevance in addition to the broader task of mapping Spitsbergen to demonstrate political control. Although the Norwegian government cut back its mining subsidies once sovereignty was secured in 1925, state investment continued where necessary to forestall Soviet acquisitions, further demonstrating the political character of mining and its associated sciences (Avango et al. 2014, 15). This in turn favored close involvement of the state as a provider not only of the stable political environment within which other activities could be conducted, but also as a direct sponsor of fieldwork, which produced artifacts such as maps that reinforced state authority (as did the presence of state-backed workers on Spitsbergen itself).

In 1928 Hoel was appointed leader of a new government institution, Norges Svalbard- og Ishavsundersøkelser (NSIU). He immediately ensured that its mandate extended to oversight of *all* expeditions to Svalbard, Norwegian, and otherwise. In a memorandum sent to "foreign powers" in 1928, and printed in the Norwegian Geographical Society's journal, the Foreign Ministry mandated that all plans for scientific work on the archipelago be submitted in advance to the NSIU (Norwegian Foreign Ministry 1928). The document may be read as an embodiment of Hoel's agenda. He and his associates would check that the planned work did not duplicate that of others, would provide safety advice, and would issue formal notification of relevant Norwegian laws (concerning the protection of animals, for instance). This was necessary "to obtain the best possible results from scientific research on Svalbard" (Norwegian Foreign Ministry 1928, 122). The fact that no fee was charged for these services only reinforced their status as naturalized aspects of the political landscape. In the years that followed, notes of thanks to Hoel and his agency appeared in print to validate the arrangement's importance, and in 1937 he proudly published a list of 25 Svalbard expeditions from nine countries to which such assistance had been given (Hoel 1937, 83–5).

Knowledge of Svalbard could be presented as the logical foundation for effective administration while signaling Norwegian commitment to the responsibilities that entailed. This could include hunting regulations and the establishment of national parks, but also producing artifacts of authority over Svalbard's geography through scientific reports and place names (Wråkberg 2002). Hoel insisted that NSIU act as a gatekeeper for proposals to name geographical features on Svalbard, providing order amidst the "chaos" produced by years of uncoordinated action (Norwegian Foreign Ministry 1928, 123). The control of names facilitated state control through the creation of a frame for administration, as well as the Norwegianization of the archipelago's geography. The name "Svalbard" could anchor such projects in the voyages of Norsemen a millennium prior. Longyear City also became Longyearbyen, Green Harbour became Grønfjorden, and so on (Hoel 1925). The project of inscribing names upon nature, thus inscribing the nation upon nature, had clear political consequences.

Another means of inscribing national order upon nature was through the creation of national parks. J.M. Coetzee has observed that wilderness may describe either the absence of naming and order—an almost pre-Adamic state of being—or a space for isolation and contemplation (Coetzee 1988, 49–50). National parks straddle the categories. Their value depends upon their assimilation into existing categories—as known spaces judged to be valuable for their assessed qualities—with instrumental value for strengthening individual or collective characters that rely upon a certain exclusion from the norm. The national park is by definition an authorized and known wilderness. As Simon Schama has asserted, "even the landscapes that we suppose to be most free of our culture may turn out, on closer inspection, to be its product" (Schama 1995, 9). Wråkberg has noted that the first major plan for nature protection on Svalbard, proposed in 1914 by the German Hugo Conwentz and supported by Norway, drew upon ideological and economic arguments rather than claims of environmental crisis (Wråkberg 2006). Nations possessed both cultural and natural heritage, the argument ran, and failure to recognize and manage either properly reflected poorly upon those in charge. This sentiment led to the formation in 1914 of Landsforeningen for Naturfredning i Norge (the National Association for Nature Protection in Norway, today known as Norges Naturvernforbund—the Norwegian Society for Conservation of Nature). Hoel joined the Association shortly after it was founded and remained an active member for over 30 years.

When it became clear in 1920 that Norway would be granted sovereignty over Spitsbergen, the Association quickly eyed up a potential new arena for action. Concluding a reflection on its recent successes—especially in protecting what he termed "nature's unique monuments"—the Association's national chairman Hjalmar Broch hoped that these developments augured well for the responsibilities Norway was about to assume in Spitsbergen (Broch 1920, 12). In May 1921 Hoel and the botanist Hanna Resvoll-Holmsen approached the Norwegian Ministry of Agriculture with suggestions for nature protection, initiating a long and not always fruitful dialogue. After 1925 the Norwegian parliament quickly passed a law spelling out how Norwegian authority would be defined and exercised. The state declared the right to administer the archipelago's flora and fauna, including a general provision for managing hunting, fishing, trapping, and other related activities—and for the "protection of animals, plants, natural formations, landscapes and archeological remains" in addition to collecting information for national statistical purposes (Stortinget 1925). The first act of protection was a ten-year moratorium on hunting the endemic reindeer sub-species *Rangifer tarandus platyrhynchus*. The decision drew heavily upon a study by zoologist Alf Wollebæk, who concluded that their range had significantly decreased due to the increased volumes and efficiency of hunting (Wollebæk 1926, 50–7). Knowledge of the reindeer and their distribution, but also of the people who had visited their habitat, were thus combined within a single framework that permitted effective administration.

Hoel described his visions for nature protection on Svalbard in some detail in the Association's annual report for 1926. The four main proposals were: to declare north-western Spitsbergen a national park; to protect the flora of another area of the island; to protect all archeological remains; and to introduce protection for both flora and fauna on the southern island of Bjørnøya (Nasjonalparker i Norge 1926, 2). In the context of this particular discussion, a nature park referred to a space in which no activities (especially economic) that impacted upon the natural environment were permitted, whereas those restrictions were not as strict in national parks, where certain areas might be preserved. Hoel's proposal for a national park on north-west Spitsbergen drew on a proposal made back in 1914—to which he had contributed—and on a narrative of control over people to facilitate control over nature (Wråkberg 2006, 14–5). The area possessed a collection of striking landscapes that in Hoel's view "represented all the more important landscape forms on Spitsbergen," comparatively unburdened by economic potential (Hoel 1926, 19). Facts concerning the precise distribution of polar bears, seals, reindeer, foxes, walruses, whales, and all manner of birds and fishes (and who had hunted them) were augmented by figures detailing how many of each animal had been hunted, and when. Arguments for hunting restrictions drew upon utility as well as sentiment (Hoel 1926, 10, 14). Effective protection measures would require constant re-evaluation and continued monitoring of animals but also of humans, including mining companies and tourists in addition to hunters. The result would be a detailed management program that Hoel could sketch almost to the last animal.

Yet the government failed to act on Hoel's proposal. Drivenes has argued that Hoel's empire at NSIU came under political attack because it was perceived as devoting too many resources to activities that did not provide direct economic benefit—such as supporting science (Drivenes 2004, 230–3). The fact that Hoel cast his argument in exactly such terms, as a means of aiding economic development in addition to legitimizing Norwegian sovereignty, could not overcome a suspicion that scientifically-informed regulations were by definition hindrances to economic growth. That argument could only be bypassed if describing Svalbard's physical geography could create a deeper emotional connection of the kind symbolized by a national park, thus constructing a natural form of cultural heritage. Works such as Resvoll-Holmsen's 1927 guide to the flora of Svalbard described the botanical phenomena of a space to which Norwegians could—and should—feel emotionally attached (Resvoll-Holmsen 1927).

Hoel assumed the Association's leadership in 1935 and almost immediately moved its administrative headquarters to NSIU's premises at the old Oslo Observatory, where NSIU office staff also took on the Association's administration. The organization that lobbied for government action on nature protection in Svalbard was now conveniently tied to the organization that oversaw state scientific research in that same space. Hoel used his position with the Association to continue to push for more nature protection on Svalbard, in addition to increasingly quixotic schemes to transplant polar fauna—culminating in an attempt to introduce penguins to continental Norway (Roberts 2011, 74–5). But attempts to declare large-scale parks on Svalbard continued to be hampered by the lack of surveillance and enforcement (Svalbard 1936, 9–10). The most notable success was the full protection from 1938 of polar bears in Kong Karl's Land, three isolated islands at the east of the archipelago. Even this decision was largely symbolic. The economic impact was minimal as the islands were hard to reach, and for those who did venture that far, hunting was not prohibited on the surrounding sea ice (Isbjørnen er fredet på Kong Karls Land 1939, 5).

The ultimate failure to implement successfully either a national park or comprehensive hunting restrictions on Svalbard was a disappointment to Hoel and the Association mainly because it reflected poorly on Norway's status as a modern, civilized nation. Failure to do so left Norway open to attacks upon the legitimacy of its rule on Svalbard, either through mismanagement of its fauna (von Staël-Holstein 1932, 18) or through failure to create national parks. One might fairly wonder whether it was even possible for Hoel to describe the archipelago and its features—animal, vegetable, and mineral—without also representing it as a Norwegian space. To him, Svalbard was a microcosm of a broader malady. His draft program for the Association's activities in 1935 led with a criticism of Norway's "extremely primitive and contradictory" nature protection laws, which he felt were inadequate compared to those of other countries (Hoel 1935). Early in his leadership a membership drive described nature protection as "a practical and sensible" means of showing love of the fatherland, and a duty to one's descendants ("Vil De bli med å verne om Norges natur?" 1935). But although Hoel and colleagues such as Gustav Smedal argued forcefully for a sense of national kinship between Norway, Svalbard, and even eastern Greenland, the polar empire they described failed to win either hearts or minds.

THE GREENING OF ARCTIC MINING LANDSCAPES

It took until 1973 before the Norwegian government established the first national parks at Svalbard. Several more have been legislated into existence since, most recently in 2002. Current environmental legislation states that "this law aims to maintain a virtually untouched environment in Svalbard with respect to continuous areas of wilderness, landscape, flora, fauna and cultural heritage" (Svalbardmiljøloven 2001, §1). The environmental law of 1973 and the national parks it established were strongly motivated by the Norwegian government's ambition to increase its influence in the archipelago, in the face of rapidly expanding oil prospecting activity in which both US and Soviet organizations were involved (Arlov 1996, 384f; see also Barr 2001, 140).

Yet, the Arctic wilderness that the environmental laws were supposed to protect also contained a significant amount of material remains from over 400 years of intensive natural resource exploitation in the archipelago. Svalbard contains the standing and ruined remains of seventeenth-century whaling stations, eighteenth-century hunting stations, nineteenth-century research stations, and most notably twentieth-century mining sites— prospecting camps, mines, transport infrastructures and mining settlements which dot the coastlines. In 1974 the Norwegian government issued legislation to recognize explicitly these human layers of the Arctic environment as cultural heritage. We argue that this cultural heritage protection legislation, as well as subsequent laws which strengthened these protection measures, should be understood as a continuation of Norway's environmental policy, and an articulation of the same motives. Cultural heritage sites are constructed to support economic and political goals in precisely the same manner as environments and natural resources. Mining companies and the tourism industry supported the Norwegian government's policies because they shared a common aim of constructing the material legacies of mining as resources for both symbolic and practical benefit.

The initial piece of cultural heritage legislation—imaginatively titled "Regulations Regarding Cultural Heritage at Svalbard and Jan Mayen," from June 1974—stated that all remains older than 1900 should be automatically protected, with the option of protecting more recent historical remains where deemed appropriate. According to Susan Barr, this law was primarily used to protect older remnants of human activity but also some industrial remains, notably in the former mining settlement of Ny-Ålesund (which was closed in the wake of an underground accident that cost 22 lives in 1962). Through the cultural heritage act of 1992, Norwegian authorities strengthened their efforts to protect the remains of human activity at Svalbard by moving the cut-off date for automatic heritage designation to 1946, with the result that many of the derelict mining sites in the archipelago now became protected as cultural heritage. The 1992 law also included the possibility to protect even younger traces of human activity as heritage (Marstrander 1999), a clause that has been exercised on a number of occasions since (most notably the vast aerial ropeway systems connecting Norwegian mines in the Advent and Longyear valleys). In 2002 the same rules were incorporated into the Norwegian Environmental Protection Act for Svalbard, a clear signal of their common purpose with other forms of environmental protection legislation.

Norwegian authorities have also contributed to the construction of industrial remains as heritage through sponsorship of the Svalbard Museum. Founded in 1979 and in 2006 re-established in new spacious facilities at the Svalbard Science Center—the largest building in Longyearbyen—the Museum is owned by a foundation, constituted by the Longyearbyen municipality, the Governor of Svalbard, Store Norske Spitsbergen Kulkompani, the University Center on Svalbard (UNIS), and the Norwegian Polar Institute. Mining has a central place in the museum's exhibition, whose narrative emphasizes the importance of mining for the twentieth-century Norwegian history of Svalbard as well as for the present day settlements, strengthening the narrative in which the material remains of mining possess value as heritage.

Another group of actors that has defined Svalbard's mining landscapes as heritage are the mining companies. Store Norske Spitsbergen Kulkompani (SNSK), the state-owned company that has been one of the main instruments of the Norwegian government strategy to maintain sovereignty at Svalbard, has built a vistor site inside a building adjacent to one of its abandoned coal mines. The company has also produced a series of lavish volumes about its own history (Westby and Amundsen 2003; Kvello 2004; Martinussen and Johnsen 2005; Holm 2006; Kvello and Johnsen 2006; Kvello 2007, 2009; Orheim 2007), as well as financially supporting research on mining history. SNSK has also provided funding for the mining exhibition at Svalbard Museum. Moreover, the company has preserved and maintained remains of their former mining operations—housing units as well as infrastructures associated with their mines.

The Russian mining company Trust Arktikugol, which has served Soviet and later Russian interest in coal and geopolitics at Svalbard since 1931, has also recently concerned itself with defining material remains of Russian activities as heritage. The company has supported the production of publications about its own history and that of other Russians on Svalbard—including the activities of North Russian Pomors, claimed to be the discoverers of the archipelago. The company is constructing a Centre for Pomor Culture in Barentsburg, the main Russian settlement. Their museum in the same town locates the Pomors as the first actors in a narrative of Russian presence on Svalbard stretching through to the present day mining activities there, articulating a long-term connection between Svalbard and Russia. Remains of Pomor presence at Spitsbergen has thereby been turned into a tool for politics (Hultgren 2002). Trust Arktikugol operates two more museums on the archipelago, in the vicinity of their abandoned mining settlements at Coles Bay and at Pyramiden.

Pyramiden provides an instructive example of Trust Arktikugol's approach to linking the industrial and the environmental. After closing down their mining operations in 1998, the company left the town to decay for a number of years. Meltwater rivers made their way into the settlement, undermining buildings, and infrastructures, while visitors from Longyearbyen looted and vandalized buildings. In 2010 the Trust—in cooperation with the governor of Svalbard—started an ambitious renovation program, and in spring 2013 the company re-opened a hotel to accommodate visitors. Trust Arktikugol has two main ambitions with this work: to develop Pyramiden as a tourist destination, with industrial heritage functioning as a material anchor for evoking a Soviet past; and to make Pyramiden into a platform for Arctic research (inspired by Ny-Ålesund), positioning Russia as a leading actor in facilitating climate change research.

A third category of actors involved in the construction of mining landscapes as heritage are companies that make up the tourist industry—a branch of the economy which on Svalbard has grown dramatically over the past two decades. Although most companies focus on selling wilderness experiences to their customers, many also sell tours with a cultural and historical focus. Guided tours at abandoned mining sites and infrastructures are offered in and around Longyearbyen, along with day trip cruises to Pyramiden, Barentsburg, and Coles Bay. At Pyramiden the tourism companies cooperate with Trust Arktikugol in selling an experience of Soviet nostalgia, the material remains of mining constructed as authentic and relics of a Soviet community designed in accordance with socialist ideals. The Soviet past that is becoming increasingly invisible in present day Russia may thus be discovered frozen in time in the Arctic. Material remains that might otherwise have been regarded as waste—as unwanted human intrusions on a pristine Arctic landscape—are thereby constructed as heritage through narratives commemorating selected aspects of the past.

The construction of mining landscapes on Svalbard as industrial heritage is thus a process that involves some of the same actors involved with environmental regulation and the protection of natural heritage. To the mining companies, material remains provide resources for legitimizing the role of mining in the history of Svalbard and for bolstering a narrative in which the activity remains a part of the archipelago's future. For much of the twentieth century, both SNSK and Trust Arktikugol did not have to worry about their futures. They could rely on firm political and financial support from their state backers, for whom the companies were instruments through which political goals could be achieved. The Norwegian state supported SNSK because the company's mines provided a platform for maintaining a Norwegian settlement, which in turn maintained the legitimacy of the Norwegian sovereignty stipulated in the Spitsbergen treaty. The Soviet state supported Trust Arktikugol because Spitsbergen coal played a role in the five-year plans of the Soviet north-west and from the beginning of the Cold War, also because the Soviet Union wanted political influence over Svalbard. The importance of Spitsbergen coal diminished in Russia after the dissolution of the Soviet Union and the new Russian government turned its political attention elsewhere. However, in recent years the Trust Arktikugol and its state owner has sought out new ways to use its abandoned mining settlements, in order to maintain its presence at Svalbard. By defining the material remains of its mining operations as heritage, the company can use them as a tool to generate alternative incomes as well as a means of performing political authority. The historical narratives which Trust Arktikugol produce seem to serve the same interest, building an image of a long and continuous Russian presence at Svalbard in which Trust Arktikugol represents the latest stage—a narrative that justifies a role for the company and its settlements in the future.

SNSK's production of history and heritage should be understood within a similar context. From the early 2000s, the Norwegian state removed much of its financial support for the company, coupled with demands that SNSK should cover its own production costs. At the same time a growing mismatch between the goals of Norway's ostentatiously progressive environmental policies and ongoing coal mining in the High Arctic (similar to its continued enthusiasm for oil) led both public opinion and decision makers to call SNSK's mining operations into question. It is clear that the company's efforts to produce history and heritage were partly, if not primarily, a way of building support for itself in the face of this new situation. By ascribing heritage status to the material remains of its operations, and locating them within the same category of valuable assets that warrant state recognition and protection, SNSK could articulate a version of its history that naturalizes its own position on Svalbard. The message is simple: SNSK made Svalbard into what it is today, and it should be a part of its future.

For the Norwegian authorities cultural heritage protection has been one among several instruments through which state authority is exercised over Svalbard. The inclusion of heritage protection in the environmental policies of the Norwegian government back in 1974 should be understood not only in the context of a global trend toward recognizing cultural heritage (most notably through the 1972 creation of the world heritage system), but also as part of a more active Norwegian Svalbard policy from the end of the 1960s (Arlov 1996). The subsequent strengthening of cultural heritage protection in 1992 and 2002 served the same purpose. Since the mid-2000s the office of the governor of Svalbard has increased its efforts to make Trust Arktikugol comply with relevant Norwegian laws. Heritage protection laws have been particularly useful, as the provision that all remains pre-dating 1946 constitute cultural heritage and are thus subject to Norwegian oversight. Pyramiden is again an instructive example. When Trust Arktikugol started to re-open this site for heritage tourism and science, the governor responded by requiring it to make an area plan. The company contracted a Norwegian architect firm for the purpose, while the governor hired heritage professionals to identify structures that came within the law's purview. Based on their report (Avango and Solnes 2013), the governor declared parts of Pyramiden as cultural heritage under Norwegian law, effectively turning parts of the town into a protected industrial heritage site. By defining material remains as heritage in need of protection, the Norwegian authorities at Svalbard are able to exercise legal authority over non-Norwegian actors and environments.

The designation and protection of industrial landscapes at Svalbard as heritage constitutes a form of politics. Current discourses about protecting cultural heritage are deeply embedded in the broader strategies of competing actors and interests, whose motives range from economic gain to the validation of national presence and the construction of mining as a legitimate activity. We are accustomed to thinking of protecting physical geographical environments (through national parks and similar instruments) as a recognition of self-evident value, but in reality the designation and protection of both natural and cultural heritage constitutes an exercise of political authority over environments.

KNOWING AND CONTROLLING SVALBARD IN THE TWENTY-FIRST CENTURY

Why are the episodes described above relevant to action in the present? First, Svalbard continues to be characterized as a space for industry, a space for wilderness, and a space for cultural heritage protection, all within an overarching theme of legitimizing the influence of competing actors. Coal mining continues to be a significant economic activity, although few believe that it will survive the current low world market prices and the position of coal in current debates on anthropogenic climate change. This final factor is particularly noteworthy. Norway has pushed hard to portray Svalbard as a space for science—most notably climate research—symbolized by its stewardship of an international research community at Ny-Ålesund, formerly a coal mining settlement. UNIS was founded in 1993 and dominates the downtown of Longyearbyen. With the strengthened environmental law of 2002, responsible environmental management has become an increasingly important aspect of Norway's claims to legitimate administration (further strengthened by its sponsorship of science). This applies also to the increasingly substantial efforts to preserve cultural heritage—including remains of mining. The power to govern Svalbard, and to determine its future, remains tied up with the production of narratives that construct the archipelago as a series of human and natural environments.

An extension of this point with particular relevance to the present is that narratives about Svalbard cannot be considered as peculiarly "Arctic" in any self-evident sense, and that the demarcation of its natural and cultural heritage reflects values from far further south. To characterize a certain space as Arctic is to incorporate it within a system of meaning that is underdetermined by physical geography. The Arctic is defined differently depending on the context, from the Arctic Circle to climatological boundaries (such as the 10 °C isotherm) to definitions based upon administrative convenience. The history of Svalbard and its representations in the form of material remains (just like its present and its future) was framed within narratives constructed elsewhere. These are Norwegian stories, Soviet stories, as well as Swedish, British, and Dutch stories, more than they are Arctic stories. While the absence of an indigenous population makes these links appear starker than they would be in Greenland, northern Canada, or Siberia, we nevertheless argue that Arctic spaces are constructed in (and often for) southern consumption, and that historiographic and analytic frames based on cartographic location must be regarded with skepticism. The growth of the Arctic as an organizing category in the twenty-first century—inscribed upon bodies such as the Arctic Council and knowledge productions such as the Arctic Climate Impact Assessment—only strengthens the need to examine critically how and why that particular category has been employed and what narratives are supported by its use (Keskitalo 2004).

Second, recognizing that describing an environment can never be separated from locating it within a political and cultural narrative is an essential prerequisite to effective advocacy. To speak of any environment, including (perhaps especially) one designated as wilderness, is to locate it within cultural and political discourses (Cronon 1995). The key insight of critical geopolitics is that the "geo" is not the space in which the political happens, but something that is created by the political to give it meaning. Consequently humanities scholars ought to ask pointedly *why* Arctic environments—natural and cultural—are constructed in particular ways, and to examine critically the narratives in which those constructions appear. Nature and heritage protection, like science or industry, involves defining environments and remains from the past rather than inscribing values upon a passive physical geography or object. It is no coincidence that the national park idea first became popular during the heyday of late-nineteenth-century nationalism. To create a reserve for the purpose of conservation or preservation is a political act that must be understood within the context of contemporaneous power relations rather than an ahistorical recognition of the ecological value of a space. Precisely the same may be said of cultural heritage.

The greatest value of the environmental humanities lies in its assertion that to speak about heritage, natural or cultural, is always to speak about people. If this can productively be done, even for a space like Svalbard, with its lack of indigenous population and relatively short history of European occupation, then rich fields await for students of other northern spaces.

WORK CITED

Arlov, Thor B. 1996. *Svalbards historie 1596–1996.* Oslo: Aschehoug.

Armiero, Marco. 2011. *A rugged nation: Mountains and the making of modern Italy.* Cambridge: White Horse Press.

Avango, Dag. 2005. *Sveagruvan: svensk gruvhantering mellan industri, diplomati och geovetenskap.* Stockholm: Jernkontoret.

Avango, Dag, and Per Högselius. 2013. Under the ice: Exploring the Arctic's energy resources, 1898–1985. In *Media and the politics of climate change. When the ice breaks,* ed. Miyase Christensen, Annika E. Nilsson, and Nina Wormbs, 128–156. New York: Palgrave MacMillan.

Avango, Dag, and Sander Solnes. 2013. *Registrering av kulturminner i Pyramiden. Registrering utfört på oppdrag fra Sysselmannen på Svalbard.* Longyearbyen: Governor of Svalbard.

Avango, Dag, Louwrens Hacquebord, and Urban Wråkberg. 2014. Industrial extraction of Arctic natural resources since the sixteenth century: Technoscience and geo-economics in the history of northern whaling and mining. *Journal of Historical Geography.* 44: 15–30. http://dx.doi.org/10.1016/j.jhg.2014.01.001.

Barr, Susan. 2001. International research in Svalbard c. 1960–1985. A cold war utopia or a pre-glasnost sparring area? In *International scientific cooperation in the Arctic,* ed. Eugene Bouzney, 96–100. Moscow: Russian Academy of Sciences.

Berg, Roald. 1995. *Norge på egen hånd 1905–1920.* Oslo: Universitetsforlaget.

———. 2004. Fornorskning av Arktis og fornorskning av Nord-Norge 1820–1920. Momenter til et helhetsperspektiv. In *Inn i riket: Svalbard, Nord-Norge og Norge,* ed. K. Zachariassen, and H. Tjelmeland, 27–38. Tromsø: University of Tromsø.

Broch, Hjalmar. 1920. Opgaver og linjer i naturfredningsarbeidet. In *Naturfredning i Norge 1920.* Kristiania: AS P.M. Bye.

Coetzee, J.M. 1988. *White writing: On the culture of letters in South Africa.* New Haven: Yale University Press.

Cronon, William. 1995. The trouble with wilderness, or getting back to the wrong nature. In *Uncommon ground: Rethinking the human place in nature,* ed. William Cronon, 69–90. New York: W.W. Norton.

Drivenes, Einar-Arne. 2004. Ishavsimperialisme. In *Norsk polarhistorie 2: vitenskapene,* ed. Einar-Arne Drivenes, and Harald D. Jølle, 175–257. Oslo: Gyldendal.

Hestmark, Geir. 2004. Kartleggerne. In *Norsk polarhistorie 2,* ed. Einar-Arne Drivenes, and Harald D. Jølle, 9–103. Oslo: Gyldendal Norsk Forlag.

Hoel, Adolf. 1925. Notes on a draft proposal prepared by the Association for the Trade Ministry regarding animal protection. National Archives of Norway,

Oslo, Naturvernforbundet collection, folder "Svalbardspørsmål: Nasjonalpark og fredning av dyr."

———. 1926. Forslag til Kongelige forskrifter vedrørende fredning, jakt, fangst og fiske på Svalbard. *Naturfredning i Norge Årsberetning.*

———. 1928. Om ordningen av de territoriale krav på Svalbard. *Norsk Geografisk Tidsskrift* 2(1): 1–24.

———. 1935. Draft proposal for the National Association for the Nature Protection of Norway annual meeting, 1935. National Archives of Norway, Naturvernforbundet collection, folder "program for 1935."

———. 1937. Report on the Activities of Norges Svalbard- og Ishavsundersøkelser 1927–1936. In *Skrifter om Svalbard og Ishavet 73.* Oslo: Jacob Dybwad.

———. 1966. *Svalbard: Svalbards historie 1596–1965,* vol 2. Oslo: Sverre Kildahls Boktrykkeri.

Holm, Arne O. 2006. *Store Norske kvinner.* Longyearbyen: Store norske Spitsbergen kulkompani.

Hultgreen, Tora. 2002. When did the pomors come to Svalbard? *Acta Borealia* 19(2): 145–165.

Isbjørnen er fredet på Kong Karls Land. 1939. In *Naturfredning i Norge Årsberetning 1938–39.* No publisher given.

Keskitalo, Carina. 2004. *Negotiating the Arctic: The construction of an international region.* New York: Routledge.

Kvello, Jan Kristoffer. 2004. *Store Norske Spitsbergen Kulkompani: Om å arbeide i en politisk bedrift på Svalbard: 1970–2000.* Longyearbyen: Store norske Spitsbergen kulkompani.

———. 2009. *Store Norske Spitsbergen Kulkompani Aktieselskap: om livet i kullgruvene på Svalbard.* Longyearbyen: Store norske Spitsbergen kulkompani.

Kvello, Jan Kristoffer, and Torbjørn Johnsen. 2006. *Store Norske Spitsbergen Kulkompani Aktieselskap : fra privat til statlig eierskap: 1945–1975.* Longyearbyen: Store norske Spitsbergen kulkompani.

Marstrander, Lyder. 1999. Svalbard cultural heritage management. In *The centennial of S.A. Andrée's north pole expedition,* ed. Urban Wråkberg. Stockholm: Royal Academy of Sciences.

Martinussen, Berit, and Torbjørn Johnsen. 2005. *Et arktisk omstillingseventyr/1987–2005.* Longyearbyen: Store norske Spitsbergen kulkompani.

Nasjonalparker i Norge: På Svalbard, Hardangervidda, Dovre og Jotunheimen. 1926. In *Naturfredning i Norge 1926.* No publisher given.

Norwegian Foreign Ministry. 1928. Om utforskningen av Svalbard. *Norsk Geografisk Tidsskrift* 2(2): 122–125.

Orheim, Olav. 2007. *Fast grunn: om bergverksordningen for Svalbard.* Longyearbyen: Store norske Spitsbergen kulkompani.

Resvoll-Holmsen, H. 1927. *Svalbards flora: med en del om dens plantevekst i nutid og fortid.* Oslo: J.W. Capellens forlag.

Riffenburgh, Beau. 1993. *The Myth of the explorer: The press, sensationalism, and geographical discovery*. London: Belhaven.

Roberts, Peder. 2011. *The European Antarctic: Science and strategy in Scandinavia and the British Empire*. New York: Palgrave Macmillan.

Roberts, Peder and Eric Paglia. 2016. Science as national belonging: The construction of Svalbard as a Norwegian space. *Social studies of science*. doi:10.1177/0306312716639153.

Schama, Simon. 1995. *Landscape and memory*. London: HarperCollins.

Spufford, Francis. 1997. *I may be some time: Ice and the English imagination*. New York: St Martin's Press.

Stortinget. 1925. Innstilling O. VIII. innstilling fra den forsterkede justitskomite om lov om Svalbard [recommendation from the strengthened judicial committee regarding the law concerning Svalbard] (Odelstinget proposition no. 48).

Svalbard. 1936. *Naturfredning i Norge Årsberetning 1936*. No publisher given.

Svalbardmiljøloven. 2001. https://lovdata.no/dokument/NL/lov/2001-06-15-79#KAPITTEL_1. Accessed 22 Sept 2016.

Toal, Gerald. 1996. *Critical geopolitics: The politics of writing global space*. Minneapolis: University of Minnesota Press.

Vil De bli med å verne om Norges natur?. 1935. National Archives of Norway, Naturvernforbundet collection, folder "program for 1935."

von Staël-Holstein, Lage F.W. 1932. *Norway in Arcticum: From Spitsbergen to—Greenland?* Copenhagen: Levin & Munksgaard.

Westby, Sigurd, and Birger Amundsen. 2003. *Store Norske Spitsbergen Kulkompani: 1916–1945*. Longyearbyen: Store norske Spitsbergen kulkompani.

Wollebæk, Alf. 1926. *The Spitsbergen Reindeer (Rangifer tarandus spetsbergensis)* NSIU Skrifter om Svalbard og Ishavet no. 4.

Wråkberg, Urban. 1999. *Vetenskapens vikingatåg: perspektiv på svensk polarforskning 1860–1930*. Stockholm: Royal Swedish Academy of Sciences.

———. 2002. The politics of naming: Contested observation and the shaping of geographical knowledge. In *Narrating the Arctic—A cultural history of Nordic scientific practices*, ed. Sverker Sörlin, and Michael Bravo, 155–198. Canton: Science History Publications.

———. 2006. Nature conservationism and the Arctic commons of Spitsbergen 1900–1920. *Acta Borealia* 23(1): 1–23.

CHAPTER 9

Toxic Blubber and Seal Skin Bikinis, or: How Green Is Greenland? Ecology in Contemporary Film and Art

Lill-Ann Körber

In a speech celebrating the implementation of self-government on 21 June 2009, Josef Motzfeldt, then president of Inatsisartut, the Parliament of Greenland, described the day as "a new start of history." In his speech, Motzfeldt made use of pertinent metaphors of nature and weather to describe the challenges of this "new era": "storms will come, we will face steep hillsides, and sometimes we will proceed on thin ice." Motzfeldt furthermore envisioned Greenland as a global player with a sustainable future: "as part of the world society, we will strive for a better future for our planet" (Motzfeldt 2009; my translation).

The Danish documentary *Greenland Year Zero* (Anders Graver and Niels Bjørn 2011) features Josef Motzfeldt as narrator. It is his voice and image we initially encounter and we are led to understand that the film's sensationalist title derives from a quotation by him. The film's title is an obvious intertext to the neorealist classic *Germany Year Zero* (Roberto

L.-A. Körber (✉)
Department of Linguistics and Scandinavian Studies, University of Oslo,
P.O. Box 1102, 0317 Oslo, Norway

© The Author(s) 2017
L.-A. Körber et al. (eds.), *Arctic Environmental Modernities*,
DOI 10.1007/978-3-319-39116-8_9

Rossellini 1948). This early post-war "rubble film" foregrounds the portrayal of building a new national imaginary on the detritus of the past, a past that nevertheless haunts the present. In a similar vein, past and present are intertwined in *Greenland Year Zero*. Interviewed at the parliament building in Nuuk, Motzfeldt addresses the many and interconnected changes—past, present, and future—facing Greenland. Political change—in the guise of self-rule—is linked with climate change and changes in economy and employment, as well as shifts in culture and identity (see Fig. 9.1). Motzfeldt is presented as a spokesperson for a self-governed and globalized Greenland in the age of global warming and the beginning of the country's exploitation of its rich natural resources.

The aim of this chapter is to reflect on the representation of Greenlandic agents and agency in recent Greenlandic art, public debate, and ecocinema (film with an explicit environmental interest, cf. MacDonald 2004). By analyzing three documentaries and an art installation, I will question how the artistic and activist contexts of these examples negotiate what Ursula K. Heise describes as a "new kind of eco-cosmopolitan environmentalism" (Heise 2008, 210). Global interdependencies require new forms of environmental awareness and ethics, a "sense of planet" which reaches beyond a "sense of place," Heise's term for an immediate connection to land often thought of as "natural" or spiritual (Heise 2008, 55). A "planetary" perspective allows for a new understanding of a community of humans and non-humans, as well as for local and global implications of environmentalism (Heise 2008, 61; see also Hennig 2014, 19–21). The recent Greenlandic eco-documentaries I emphasize in this chapter point to an eco-cosmopolitan understanding of environmentalism. The realization of Greenland's position in a globalized world implicates new figurations of identity and territory, where the local and the global are not necessarily juxtaposed, but intertwined. However, the selected examples

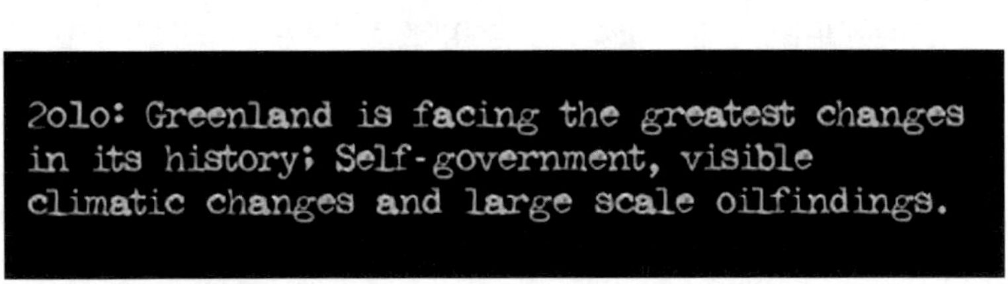

2o1o: Greenland is facing the greatest changes in its history; Self-government, visible climatic changes and large scale oilfindings.

Fig. 9.1 Still from *Greenland Year Zero* (2011). Courtesy Anders Graver, Humbug Film

of Greenlandic and Greenland-related art and film point at the pitfalls of a "sense of planet" if it does not acknowledge a "sense of place."

ECOLOGICAL DISCOURSES IN AND ABOUT GREENLAND

Ecology in Greenland is globalized in two senses. Through ecological systems and climate change, it is connected to the rest of the world, and directly affected with respect to global warming, natural resources, and pollution of oceans and the atmosphere. The same interconnectedness is true for the level of ecological discourses, practices, and representations. The situation in Greenland is specific, however, in the sense that 57,000 Greenlanders are confronted with worldwide attention aimed at their precarious surroundings, for instance with—as of January 2016—more than 7.5 million signatories for Greenpeace's "Save the Arctic" campaign, of which the vast majority live far from the Arctic. The signatories might be aware of the global implications of ecological changes in the Arctic, but lack a "sense of place," that is knowledge of local living conditions. The Greenpeace campaign's setup in fact diminishes the voices of Arctic residents, as Mered has argued (2013). Moreover, such disenfranchisement of local perspectives has a long history.

Activism against Greenpeace began to form in Greenland in the 1970s when the campaign against industrialized seal hunting—which has never been practiced in Greenland—resulted in a ban on the import and export of seal products. This ban has severe effects on the sustainably operating Greenlandic hunters, up to the present day, a fact that has not been forgotten by many Greenlanders who remain skeptical about the presence and interference of international environmental activists (Hauptmann 2014b and the Facebook group "Greenpeace out of Greenlandic territories"). The pressing issue concerning the agency and sovereignty of interpretation of Greenlanders and other Arctic residents arises not only in the context of environmental activism, but also with regard to media and artistic representations of ecological changes, knowledge, and expertise. Recent critical journalism and scholarship has pointed to power relations intrinsic to the field of Arctic ecologies—among other things to the construction of "experts," most of whom are not Arctic residents, and to the establishment of discourses, metaphors, and narratives in service of such "expert" perspectives (Bjørst 2014; Nuttall 2012; Thórsson 2014). These discourses, metaphors, and narratives are not mimetic representations of natural phenomena. Rather, they have served, since the era of

Arctic exploration, as iconographies and representational and analytical modes. They charge Arctic nature and ecology with a meaning that potentially tells us more about the authors' aesthetic, economic, and ideological interests than about what is depicted. To complicate the picture further, it seems necessary to acknowledge the existence of multiple and potentially conflicting alternative (local) narratives.

Biologist and writer Aviaja Lyberth Hauptmann, who runs a blog on Greenland's ecological development, points to the diversity of Greenlandic voices due, among other things, to differences with respect to education, occupation, generation, language proficiency, place of residency, and social and material welfare. Roughly, the population in south-western Greenland live in bigger cities and have a higher than average level of education; there is also a larger number of Danish-, bi-, or multilingual speakers. The north and east are less urbanized and home to comparatively more monolingual Greenlandic-speakers, some of whom still pursue traditional occupations. Thus, a "sense of place" has perhaps mostly been developed and discussed in Greenlandic, while a "sense of planet," in the sense of transnational environmental and scientific discourses, is being debated in Danish or English in the bigger cities. Regardless of their level of (science-based, "Western") knowledge, Hauptmann maintains, Greenlanders tend to be reserved towards outsiders, including scientists and politicians who determine fishing quotas or generally want to have a say in the Greenlanders' utilization of their natural resources (Hauptmann 2012, 2014a, b). Protests against Greenpeace's anti-oil drilling campaigning in 2009 in Nuuk were in equal shares motivated by the prospect of the economic and ultimately political benefits of industrialization, and by resentments against interference in domestic matters.

Yet not everybody in Greenland favors heavy industry and mining either, and there are weighty Greenlandic actors in the fields of environmental awareness and activism, among them the Inuit Circumpolar Council (ICC) and Minik Thorleif Rosing, a geologist at the Danish Natural History Museum and the University of Copenhagen. For instance, in cooperation with Kalaallit Nunaanni Aalisartut Piniartullu Kattuffiat (The Association of Fishermen and Hunters in Greenland), ICC Greenland conducted its own study of the impacts of climate change in the country, the Sila-Inuk project (2005–10), with the goal of collecting climate change observations made by residents (inuit.org; Holm 2010). As a renowned scientist, Rosing is a public voice in the debate about Arctic and more specifically about Greenlandic ecology. Born in Greenland, where he spent his childhood, and educated in Denmark and the United States, he can be said to represent, or embody, both paradigms of ecological knowledge. Rosing has both local experience and is engaged in global ecological discourses and scientific methodology. It is perhaps this double "expertise," besides his communication talents and ambitions, which makes him a sought-after figure in public debates, the media, and art projects. Most recently he participated in the expedition and ensuing film *Expedition to the End of the World* (Daniel Dencik 2013) and initiated, together with Icelandic–Danish artist Ólafur Elíasson, the installation project *Ice Watch* on the occasion of COP21, the climate summit in Paris in December 2015.

Ecological activism can thus be said to take place at the intersection of politics, science, and art. How, then, is the interconnectedness of a "sense of place" and a "sense of planet," the relation between the local and the global in Greenlandic ecological discourse, reflected in recent films? What the selected examples have in common is that they powerfully counter the widespread tendency of environmental documentaries from outside the Arctic to imagine the place as a desolate blank space, without residents, or to situate them as witnesses or victims anchored in traditional lifestyles in ways that seldom present them as active, cosmopolitan, mobile, or educated actors. In short, Arctic residents are rarely presented as "experts." So who are these eco-cosmopolitan Greenlanders? Which filmic means are used to address and negotiate the question of agency and its constitution? Which alternative modes of interpretation and explanation, of strategies with respect to ecological challenges, do the films present?

GLOBALIZED AND INDUSTRIALIZED? *GREENLAND YEAR ZERO*

Greenland Year Zero, a Danish production screened at international documentary film festivals and cultural institutions in Denmark, first introduces politician Josef Motzfeldt as a wise elder, soon to hand over the country's fate to a younger generation. The film then presents four Greenlandic teenagers from Nuuk and Aasiaat. Shown through close-ups, we get glimpses of their everyday life, accompanied by short comments about their present and future. The interviews were conducted in Danish, which all interviewees spoke; they are presented as part of the so-called "self-rule generation": urban, cosmopolitan, bi- or multilingual, and, as students of one of the four high schools in Greenland, well-educated. The film juxtaposes close-ups and fixed camera shots. This technique produces a number of tableaux that juxtapose humans with nature. It is important, in the context of Arctic environmental imagery, that the film emphasizes cityscapes, cultivated and industrial landscapes, busy construction sites, or tamed nature in the context of man-made structures (Fig. 9.2 and 9.3).

Today's Greenland, according to *Greenland Year Zero*, is a globalized industrial and information society. We see corporate facilities such as the freight and transportation company Royal Arctic, seafood producer Royal Greenland, and Scottish oil drilling company Cairn Energy. These images present a scenario of continuous utilization of natural resources, as well as a new place of heavy industrialization (the high expectations of the years around 2009 of lucrative extraction of fossil fuels and ensuing economic autonomy have since subsided). In one of his comments, Motzfeldt criticizes the outside world for preferring Greenland to remain a society of hunters, whalers, and fishermen, while restricting, or banning, the trade with seal and whale products. In *Greenland Year Zero*, the relation of humanity and nature, of identity and place, is neither romanticized nor symbolic. The film instead foregrounds a symbiotic coexistence in accordance with the needs of an autonomous, educated, internationally oriented, and highly mobile population with globalized consumption habits.

Ecological hazards such as melting ice or a risk of oil disasters are mentioned by some of the young people in the film. What is more, in one scene, Jonas, from the western Greenlandic town Aasiaat, watches the coverage of a Greenpeace campaign against oil drilling offshore on a laptop at school.

Fig. 9.2 Still from *Greenland Year Zero* (2011). Courtesy Anders Graver, Humbug Film

Fig. 9.3 Photograph from the making of *Greenland Year Zero* (2011). Courtesy Anders Graver, Humbug Film

The journalist's voice is part of the film's soundscape, mixed with the sounds of wind, water, and urban life. The English-speaking voice is used as an ethereal voice-over, representing an international concern for ecological consequences of global warming and the exploitation of natural resources in Greenland and in the Arctic. On the solid ground of southern and western urban Greenland, and on the level of the film's imagery, this concern is, however, aligned with the perhaps more urgent concern for a sustainable future for an autonomous, or even sovereign, Greenlandic society, a perspective that includes ecological, economic, educational, and political considerations.

Local Perspectives on Climate Change: *Green Land*

Green Land (*Nuna Qorsooqqittoq/Grøn Land*, 2009, see Fig. 9.4) is a documentary directed by Aká Hansen in the context of the most productive film company of recent years in Greenland, Tumit Production. It premiered in the context of the COP15 climate summit in Copenhagen and was shown in culture centers in Greenland and Denmark and at several Nordic film festivals before it was made publicly accessible on YouTube. Tumit Production's explicit *raison d'être* of recent years has been to pro-

"It's so nice weather all the time."

Fig. 9.4 Still from *Green Land/Nuna Qorsooqqittoq/Grøn Land* (2013). Courtesy Aká Hansen, UILU Stories

vide Greenlanders with home grown entertainment, especially feature films in several popular genres such as comedy, horror, or thriller. Hansen, who has since moved back to Denmark and founded her own production company UILU Stories, has been one of the most outspoken advocates for a self-conscious, self-rule generation explicitly positioning themselves beyond the mental restrictions of postcolonialism (for an overview of this situation in Greenland, see Körber and Volquardsen 2014; Pedersen 2014; Thisted 2014). Within the emerging film scene, including the Greenland Association of Film Workers FILM.GL and the Greenland Eyes International Film Festival (2012–15), Hansen and other contemporary Greenlandic filmmakers are simultaneously locally and globally oriented creative artists. Moving between Denmark, Greenland, and other locations, they act locally with respect to intended audience and subject matter and globally in terms of cultural influences and references. *Green Land* is the only "proper" Greenlandic example of films discussed in this chapter, and it is the only one with its main focus on climate change.

Presenting exclusively local perspectives on the phenomenon, the film foregrounds possible consequences for Greenland, its inhabitants, and its flora and fauna (see Fig. 9.5). Similar to *Greenland Year Zero*, it is based on the points of view of five young Greenlanders. We get to know their lives and reflections about climate change in several rounds of comments, with interludes of images of their home environments. Again, these surroundings consist less of the pristine wilderness represented in

There are good opportunities to grow vegetables.

Fig. 9.5 *Still from* Green Land/Nuna Qorsooqqittoq/Grøn Land *(2013). Courtesy Aká Hansen, UILU Stories*

traditional outside Arctic imagery, but instead more of urban landscapes, industrial sites, or vistas from the perspective of the interviewees and their homes. The film presents hybrid and vernacular spaces of Arctic urbanism. In contrast to *Greenland Year Zero*, however, the young people in *Green Land* speak Greenlandic and the interviews have obviously been conducted in their first language. The choice of language reflects both the intended audience and recent linguistic developments in Greenland. *Green Land* is a film first and foremost by and for Greenlanders, representing local discourses. What is more, since the implementation of a new language law in 2009 as part of the larger shifts in the context of self-government and extended autonomy from Denmark, Greenlandic is the sole official language, and has rapidly become predominant in public, administrative, and cultural discourse (a fact that puts pressure on Danish speakers, reversing earlier power structures linked to language in Greenland).

Compared to *Greenland Year Zero*, *Green Land* focuses in more detail on the interviewees' living conditions and realms of experience. Two different sources of knowledge about climate change thus become apparent: one from international discourses that the five learn about at school and in the media (Fig. 9.6), another from individual and collective memory. They mention that there is less snow today than during their childhood, that certain animals or species have appeared or disappeared (especially insects), and that the behavior of fauna has changed. Dimensions of daily

Not until our teachers in high school told us about global warming –

Fig. 9.6 Still from *Green Land/Nuna Qorsooqqittoq/Grøn Land* (2013). Courtesy Aká Hansen, UILU Stories

life and personal experience are at the forefront and the interviewees, while well informed and globally conscious, emphasize a pragmatic approach to ecological changes and the benefits of a warmer climate. One of the interviewees studies agriculture, another one is a sheep farming apprentice, a third one a sheep farmer's daughter, and a fourth works for Air Greenland. These occupational fields would all benefit from a longer snow and ice-free season. The interviewees make suggestions for more environmentally aware behavior such as saving water, electricity, and fuel. As the film was shot during summer in the more densely populated south-west of Greenland, we see glaciers on the edge of the icecap, ice-free coastal areas, cultivated landscapes, grass, flocks of birds, and domestic animals such as sheep. We thus literally see a green Greenland (Fig. 9.7 and 9.8).

The perspectives of *Green Land* coincide with the research findings of anthropologist Mark Nuttall and Arcticist Lill Rastad Bjørst about perceptions of climate change in Greenland (Nuttall 2010, 2012; Bjørst 2011, 2012, 2014). They have noted reserved attitudes towards scientific findings and instead find a trust in local methods, traditions and practices of observation, adaption, and anticipation. What is more, they have noticed in Greenlandic discourses about climate change a predominance of notions of continuity and cyclical development rather than linear narratives and concepts such as the widely used metaphors of "tipping points," crisis, and catastrophe.

Fig. 9.7 Still from *Green Land/Nuna Qorsooqqittoq/Grøn Land* (2013). Courtesy Aká Hansen, UILU Stories

Fig. 9.8 Still from *Green Land/Nuna Qorsooqqittoq/Grøn Land* (2013). Courtesy Aká Hansen, UILU Stories

INTERCONNECTED GREENLAND: TRANSNATIONAL ACTIVISM IN *SILENT SNOW*

As an example of ecocinema from and about Greenland, *Silent Snow* (2011) emphasizes an explicit eco-juristic, activist agenda. Filmed entirely in English, *Silent Snow* is aimed at an international audience. The film has been featured at many environmental and indigenous film festivals around the globe, won several awards, and was supported by Greenpeace and other environmentally oriented NGOs and institutions. The narrator and subject of the investigation is Pipaluk de Groot (at the time of the film shooting, Pipaluk Knudsen-Ostermann), a Greenlander who now resides in the Netherlands. Her language and intercultural communication skills make her an eco-cosmopolitan role model par excellence (Fig. 9.9).

The main point of the film is the juxtaposition of local and global perspectives, of a "sense of place" with a "sense of the planetary." The narrative and spatial point of departure, which helps establish the film's narrative frame, is Pipaluk de Groot's northern Greenlandic hometown Uummannaq. The film continues as a journey by icebreaker and dog sled through the polar sea to a village further up north. The journey documents de Groot and her family and friends' concerns about melting ice and, above all, the pollution of snow, ice, and the local maritime fauna and traditional food source—seals, fish, and whales—by pesticides emitted elsewhere in the world. The concerns are articulated in voice-over and dinner-table conversations over a meal of *matak* (blubber), illustrating the direct impact of the invisible maritime pollution on peoples' lives; although produced and utilized in distant parts of the world, the concentration of pesticides in marine mammals, the main diet of residents in the non-arable High Arctic for thousands of years, has reached a dangerous level, especially for children and pregnant women. During the course of the film, we follow de Groot on her quest to trace the sources of this pollution. A global network of victims of ruthless corporations and corrupt governments, indigenous peoples, threatened landscapes, and unborn babies unfolds, and Greenland, far from being remote and peripheral, is directly connected with pesticide producers and users in East Africa, Costa Rica, and India (Fig. 9.10).

The outlook of *Silent Snow* on environmental changes and challenges in Greenland is unambiguous: "Greenland is in the news a lot, and it is bad news," says de Groot at the beginning of the film, reflecting the fact that the country, with its iconic polar bears, ice bergs, and ice fjords, has represented the center of a world map of anthropogenic climate change in recent years.

Fig. 9.9 Still from *Silent Snow* (2011). Courtesy Jan van den Berg

Fig. 9.10   Still from *Silent Snow* (2011). Courtesy Jan van den Berg

Former Greenlandic Minister of Culture, Education, Science, and Religion Henriette Rasmussen expresses her concern that "modest consumers" like Greenland fall victim to large-scale polluters elsewhere. As a consequence, the film's quest seeks, and finds, solidarity and allies among the underprivileged and silenced. De Groot feels and establishes "a strong connection" with the Masai of Tanzania via their and the Inuit's shared preference for unsalted meat. She identifies with the BriBri of western Costa Rica on the basis of their use of natural resources: "they only take what they need. They don't exploit. … We share the same respect" (Fig. 9.11).

In order to further its cause, *Silent Snow* draws on images of unspoilt nature, purity, and authenticity (Fig. 9.12). At the same time, the film uses what Paula Willoquet-Maricondi and Jennifer Marchiolatti have described as the main features of "indigenous ecocinema" (2010; see also Marchiolatti 2010): the presentation of alternative worldviews, relationships, and a spiritually charged connection or interdependence between humanity and nature. Indeed, besides—or overlapping with—current trends to connect Greenland to globalized, mainly US, popular culture, there is a growing interest among young Greenlanders to revitalize symbols, practices, and spirituality of the circumpolar Inuit (Körber 2014; Rossen forthcoming; Thisted 2015 and chapter "The Greenlandic Reconciliation Commission: Ethnonationalism, Arctic Resources, and Post-Colonial Identity" in this volume; also see the work of tattoo artist Maya Sialuk Jacobsen).

Fig. 9.11 Still from *Silent Snow* (2011). Courtesy Jan van den Berg

Fig. 9.12 Still from *Silent Snow* (2011). Courtesy Jan van den Berg

A Seal Skin Bikini Hung to Dry, or: Can You Make Jokes About Climate Change?

Greenlandic images of ecological debates also contain another, and often missing, mode of negotiating these issues: humor. This mode is one not often attributed to the alleged victims of misrepresentation (Thisted 2006), but is obviously present as part of recent Greenlandic art, culture, and activism. Bolatta Silis Høegh's art installation entitled *Allotment Society "Sisimiut" in the Year 2068* (*Haveforeningen "Sisimiut" Anno 2068*) was included in two group exhibitions. It was first exhibited as part of the show "In the Eye of Climate Change" at Nordatlantens Brygge (the North Atlantic Wharf) in Copenhagen in 2009 in the context of the COP15 climate summit. Later it was included in "KUUK," a collection of Greenland related art, curated by the artists, critics, and writers Iben Mondrup and Julie Edel Hardenberg, and exhibited in Nuuk and Copenhagen in 2010. In the context of the latter exhibition, Høegh received an award from the Danish Arts Foundation for her work (Fig. 9.13).

Høegh's installation reflects a merger of perspectives on climate change and of contemporary Greenlandic culture, art, and political debates. What we see is a tiny lush garden plot with tropical vegetation such as coconut palms. Scattered around the lawn are a Greenlandic sledge, an issue of the magazine *Greenland Today*, advertising the Olympic summer games in Greenland in the year 2072, and a traditional women's knife, the *ulo*. On a laundry line hangs a bikini to dry (Fig. 9.14). Made from seal skin and accompanied by a short-sleeved anorak (in its long-sleeved version part of the national costume for men), these garments provide a future vision of the effects of global warming. The artist added in an interview that, in contrast to its original functions of dog sledding and the flaying and partition of seals, the sled is supposed to be used as a sun bed and the *ulo* to crack open the coconuts. In Høegh's rendition, Sisimiut, a city in western Greenland close to the Arctic circle, will in 60 years time have metamorphosed into a Dano-Greenlandic tropical paradise, a hybrid of Inuit culture, Danish summer houses, and clichés of tropical tourism. Training fields for sled dogs will have to make way for the summer fantasy of urban dwellers.

Høegh's project is echoed by the public Facebook event "Greenland Beach Party 2032!" According to the announcement, the party will take place on 16 July 2032, from 2.00 to 11.30 p.m., and the motto (and only information) is "Hallelujah to Global Warming." In her art piece,

Fig. 9.13 Bolatta Silis Høegh, *Haveforeningen "Sisimiut" anno 2068* (2010). Installation at the exhibition In the Eye of Climate Change at Nordatlantens Brygge, Copenhagen. Photograph by Ivars Silis. Courtesy the artist

Høegh chooses a humoristic angle on the narrative of crisis and catastrophe characteristic of representations of climate change. Her art work plays with many heteroimages of Greenlandic culture and nature and starts with a joke: What if Greenland really was green, not in the way south Greenland's flora is green, or green in the sense of the Viking name givers (the Greenlandic/Kalallissut word for the country does not contain the color), but in the sense of tropical greenery? If it simply refuses to be the blank spot on the map, a place of eternal snow and ice, devoid of humans except traditionally clad hunters? Moreover, the project hints at the story that Greenland's national soccer team is not accredited by FIFA, not only because the country is not a sovereign nation state, but because of the lack of grass soccer fields. All aspects of this imagined change concern the relation of collective identity and territory insofar as the influence of climate on mentality is a powerful, persistent, as well as banal, narrative. The Greenlanders lounging on their sled sun bed in their seal skin bikinis neither want nor need to be saved: they are adaptable, they are having fun, and they are making fun of themselves. The only thing that is saved here is money: as one participant writes on the beach party's Facebook page: "This will be epic! We have 20 years to save for flights and booze!".

Fig. 9.14 Bolatta Silis Høegh, *Haveforeningen "Sisimiut" anno 2068* (2010). Photograph by Ivars Silis. Courtesy the artist

To do justice to the complexity of Bolatta Silis Høegh's work, one must add that her interpretation of yet another potential ecological challenge in Greenland, namely the extraction of uranium, turned out to be much less humorous. She produced a series of large paintings as part of a physical reaction to the controversial decision of the Greenlandic government, Naalakkersuisut, to lift the ban on uranium extraction, known as zero tolerance policy, during fall 2013. The paintings were subsequently exhibited in Copenhagen and Narsaq under the titles "Lights On Lights Off" and "STORM" (Nordatlantens Brygge, Copenhagen, 2015/16) and link the unsustainable and potentially lethally dangerous exploitation of natural resources to injured naked bodies. The violated bodies, some of them self-portraits, can be read as allegories of scarred land, left to bleed by shortsighted interventions in the precarious conditions of the human and non-human, of real, mental, and emotional landscapes (cf. Norman 2014, 2015).

Conclusion: New Ecocritical Perspectives in Greenlandic Cinema and Art

Despite their differences in content and form, the films *Greenland Year Zero*, *Green Land*, and *Silent Snow*, the art work *Allotment Society "Sisimiut" in the Year 2068*, and the Facebook happening "Greenlandic Beach Party 2032!" all point to the emergence of new types of representations addressing Greenlandic politics and environmental issues. The films contrast with earlier mainstream environmentalist cinema; in these, nature is neither presented as pristine wilderness nor as uninhabitable or untamable. Instead, the environment emerges—in different ways and to different degrees—as man-made, connecting current Greenlandic discourses with the idea of the Anthropocene. Traditional lifestyles have become a leisure activity, a commodity, or a sideline to people's main occupation. This reflects the fact that international legislation, such as fishing quotas or bans on local produce—some of it paradoxically established under pressure of environmental activism—has had an ambivalent foundational impact on the modernization of Greenland (cf. chapter "Cod Society: The Technopolitics of Modern Greenland" in this volume). The films make use of landscape imagery mainly as a context for human activity and in the context of everyday life—as vernacular landscapes as opposed to the sublime landscapes that figure prominently in traditional representations of the Arctic. *Greenland Year Zero* and *Green Land*, as well as Høegh's art, avoid the use of what Bjørst describes as icons of the international ecological discourse about climate change and the Arctic: polar bears and hunters (Bjørst 2014). These renditions rather support Nuttall's observation that "climate change does not necessarily threaten Greenlanders, but empowers them" (Nuttall 2010, 29).

In contrast to other regions in the world, the changes of climate, flora, and fauna are real and palpable for most Greenlanders. But according to Nuttall's findings, there is a widely spread trust that the changes can be managed with local ecological knowledge; in the words of Greenlandic climate scientist H. C. Petersen: TEK (traditional ecological knowledge). The potentially patronizing framework of an understanding of indigenous peoples as witnesses and above all victims of global developments only partly apply to Greenland—as conveyed through the films and art works discussed in this chapter—because Greenlanders form the majority in their own self-governed country and, despite economic and social challenges, boast a well-educated, well-informed, and mobile young generation. This is especially clearly formulated in *Green Land* and *Greenland Year Zero*, which turn their attention to the more urbanized south-western regions. These films communicate a critical eco-cosmopolitanism that challenges an understanding of universal ecological principles. In these examples of ecocinema, which reflect the political and economic context of today's Greenland, the issue of environmental justice is weighed against international law and national self-determination with regard to the sovereignty over natural and human resources. The films abstain from both eco-romanticism and apocalyptical visions, or use them strategically, and in so doing implicitly and critically refer to traditional representations of Greenland and the Arctic. In all these examples, we find a balance between local and global implications of the environmental issues in question, between "a sense of place" and "a sense of planet."

The particular situatedness of most recent Greenlandic ecocritical filmmaking vis-à-vis ecocinema and indigenous cinema and native filmmaking approaches is the simultaneity of the evolving film scene and introduction of self-rule, and therefore of political, economic, and climate change. Ecocriticism with regard to Greenland must take account of these specific historical and present developments. In contrast to the agendas of ecocriticism and ecocinema elsewhere, it is not necessarily a shift from anthropocentrism to ecocentrism that is at stake here, but rather the opposite. To acknowledge an anthropocentric Greenland is also an acknowledgment of its sovereignty, agency, and humor.

Acknowledgments Thanks to Christina Just, Jan van den Berg, Anders Graver, Pipaluk de Groot, Aviaja Lyberth Hauptmann, Aká Hansen, and Bolatta Silis Høegh for support and helpful comments.

ART WORKS

Bolatta Silis Høegh: *Haveforeningen "Sisimiut"Anno 2068* (Allotment Society "Sisimiut" in the Year 2068, 2009/10).

Greenland Year Zero (Anders Graver, Niels Bjørn, DK 2011, 26 min): www.imdb.com/video/wab/vi3854999065

Green Land/Nuna Qorsooqqittoq/Grøn Land (Aká Hansen/Tumit Production, GL 2009, 24 min): www.youtube.com/watch?v=9bTuZNxv Gj4&index=2&list=PLJ5eAl_e_VbzuSz-_S3WfMpp8ybY8upZJ

Silent Snow (Jan van den Berg, Pipaluk Knudsen-Ostermann, NL 2011, 71 min)

WORK CITED

Bjørst, Lill Rastad. 2011. Klima som sila. Lokale klimateorier fra Diskobugten. *Tidsskriftet Antropologi* 64: 89–99.

———. 2012. Climate testimonies and climatecrisis narratives. Inuit delegated to speak on behalf of the climate. *Acta Borealia* 29(1): 98–113.

———. 2014. Arktis som budbringer. Isbjørne og mennesker i den internationale klimadebat. In *Klima og mennesker. Humanistiske perspektiver på klimaforandringer*, ed. Mikkel Sørensen, and Mikkel Fugl Eskjær, 125–144. Copenhagen: Museum Tusculanum.

Ekspeditionen til verdens ende (Expedition to the End of the World, Daniel Dencik, DK 2013, 90 min).

Germania anno zero (Germany Year Zero, Roberto Rossellini, I 1948, 78 min).

Hauptmann, Aviaja Lyberth. 2012. Hvad er grønlændernes holdning til storindustriens invasion? *Grønlandsbloggen*, March 15, 2012. http://ing.dk/blog/hvad-er-groenlaendernes-holdning-til-storindustriens-invasion-127670

———. 2014a. Støt de grønlandske sælfangere! *Grønlandsbloggen*, March 13, 2014. http://ing.dk/blog/stoet-de-groenlandske-saelfangere-167030

———. 2014b. Private email correspondence, October 2014.

Heise, Ursula K. 2008. *Sense of place and sense of planet: The environmental imagination of the global*. New York: Oxford University Press.

———. 2011. Developing a sense of planet: Ecocriticism and globalisation. In *Teaching ecocriticism and green cultural studies*, ed. Greg Garrard, 90–103. Basingstoke: Palgrave Macmillan.

Hennig, Reinhard. 2014. *Umwelt-engagierte Literatur aus Island und Norwegen. Ein interdisziplinärer Beitrag zu den environmental humanities.* Frankfurt a.M.: Peter Lang.

Holm, Lene Kielsen. 2010. Sila-Inuk. Study of the impacts of climate change in Greenland. In *SIKU: knowing our ice. Documenting Inuit Sea Ice knowledge and use*, ed. Igor Krupnik et al., 145–160. Berlin: Springer.

ICC Greenland. Sila-Inuk: A study of the impact of climate change in Greenland. http://inuit.org/en/climate-change/sila-inuk-a-study-of-the-impacts-of-climate-change-in-greenland.html

Körber, Lill-Ann. 2014. Mapping Greenland: The Greenlandic flag and critical cartography in literature, art and fashion. In *The postcolonial North Atlantic: Iceland, Greenland and the Faroe Islands*, ed. Lill-Ann Körber, and Ebbe Volquardsen, 361–390. Berlin: Nordeuropa-Institut der Humboldt-Universität zu Berlin.

Körber, Lill-Ann, and Ebbe Volquardsen. 2014. *The postcolonial North Atlantic: Iceland, Greenland and the Faroe Islands.* Berlin: Nordeuropa-Institut der Humboldt-Universität zu Berlin.

MacDonald, Scott. 2004. Toward an eco-cinema. *Interdisciplinary Studies in Literature and Environment* 11(2): 107–132.

Machiorlatti, Jennifer A. 2010. Ecocinema, ecojustice, and indigenous world-views: Native and first nations media as cultural recovery. In *Framing the world. Explorations in ecocriticism and film*, ed. Paula Willoquet-Maricondi, 62–80. Charlottesville/London: University of Virginia Press.

Mered, Mikå. 2013. Greenpeace in the Arctic. Activists or pirates?. *The Arctic Journal*, October 4, 2013. http://www.thearcticjournal.com/opinion/157/greenpeace-arctic-activists-or-pirates

Motzfeldt, Josef. 2009. Tale af Josef Motzfeldt om indførelse af Selvstyret, June 21, 2009. http://www.inatsisartut.gl/media/17530/Tale%20af%20Josef%20Motzfeldt%20om%20indførelse%20af%20Selvstyret%20DA.pdf

Norman, David Winfield. 2014. Preface. Exhibition catalogue *Bolatta Silis Høegh: Lights on, Lights off.* Published on the occasion of a solo exhibition at Kongelejligheden in Kastrup, Denmark, and Nuuk Kunstmuseum, Greenland, funded by NAPA. http://bolatta.com/about

———. 2015. Naturkraft. Exhibition catalogue *Bolatta Silis Høegh: STORM.* Copenhagen: Nordatlantens Brygge.

Nuttall, Mark. 2010. Anticipation, climate change, and movement in Greenland. *Études/Inuit/Studies* 34(1): 21–37.

———. 2012. Tipping points and the human world: Living with change and thinking about the future. *AMBIO* 41(1): 96–105.

Pedersen, Birgit Kleist. 2014. Greenlandic images and the post-colonial. Is it such a big deal after all? In *The postcolonial North Atlantic: Iceland, Greenland and the Faroe Islands*, ed. Lill-Ann Körber, and Ebbe Volquardsen, 283–311. Berlin: Nordeuropa-Institut der Humboldt-Universität zu Berlin.

Rossen, Rosannguaq. Forthcoming. Nationbranding i Grønland – set igennem mode. *Grønlandsk kultur- og samfundsforskning* 2016. Nuuk: Ilisimatusarfik/ Forlaget Atuagkat.

Thisted, Kirsten. 2006. Eskimoeksotisme – et kritisk essay om repræsentationsanalyse. In *Jagten på det eksotiske*, ed. Lene Bull Christiansen, 61–77. Roskilde Universitet: Institut for Kultur og Identitet.

———. 2014. Imperial ghosts in the North Atlantic: Old and new narratives about the colonial relations between Greenland and Denmark. In *(Post-) Colonialism across Europe: Transcultural history and national memory*, ed. Dirk Göttsche, and Axel Dunker, 107–134. Bielefeld: Aisthetis Verlag.

———. 2015. Cosmopolitan Inuit: New perspectives on Greenlandic film. In *Films on ice: Cinemas of the Arctic*, ed. Scott MacKenzie, and Anna Westerståhl Stenport, 97–104. Edinburgh: Edinburgh University Press.

Thórsson, Elías. 2014. An avoidable truth: We can't save the climate. *The Arctic Journal*, May 15, 2014. http://arcticjournal.com/climate/613/avoidable-truth-we-cant-save-climate

Willoquet-Maricondi, Paula (ed). 2010. *Framing the world. Explorations in ecocriticism and film*. Charlottesville/London: University of Virginia Press.

CHAPTER 10

The Negative Space in the National Imagination: Russia and the Arctic

Lilya Kaganovsky

Russia is one of five countries bordering the Arctic Ocean, with about one-fifth of its current landmass north of the Arctic Circle and the longest coastline of all circumpolar nations. In 2011, out of 4 million inhabitants of the Arctic, roughly 2 million lived in Arctic Russia, making it the largest Arctic country by population. Indeed, as Dominic Basulto has noted, to understand properly what the Arctic means to Russia (or, as he puts it, what "Russia is up to in the Arctic"), you will need to throw out your atlases and your Mercator projection maps of the world (Basulto 2015). You'll need to delete Google Maps and Apple Maps from your smartphone. Instead, what you'll need to do is pull out another Mercator map—the famous "Septentrionalium Terrarum descriptio" of 1595—considered by cartographers to be the first ever dedicated map of the Arctic. This gorgeous map is the key to understanding Russia's current Arctic strategy. "Once you get used to viewing the world from the admittedly disorienting perspective of the North Pole," writes Basulto,

> you'll notice that there are a few oddities here—the inscription that a band of female pygmies inhabit an outlying island of Norway, the vast whirlpool and rivers at the top of the world, or the black magnetic mountain at the North Pole.

L. Kaganovsky (✉)
Associate Professor of Slavic, Comparative Literature, and Media and Cinema Studies University of Illinois, Urbana-Champaign, USA

L.-A. Körber et al. (eds.), *Arctic Environmental Modernities*,
DOI 10.1007/978-3-319-39116-8_10

You can see at a glance that not only is Russia a hulking Eurasian land-mass, but that it is also potentially a huge Arctic superpower. Check out the breadth and expanse—it's almost like a leviathan of the High North extending from Scandinavia to the Bering Strait. The only other countries that come close to Russia in size are Canada, Norway, and Denmark (by virtue of its claim to Greenland). These four nations all dwarf the US Arctic landmass (i.e., Alaska). (Basulto 2015)

The Russian "North"—officially defined as a region in 1960, and which encompasses Arctic Russia and Siberia, as well as territory from St Petersburg to the Far East—covers approximately 70 % of Russia's total land area, but contains just 7.9 % of its population. At 11.9 million sq. km, it would be the world's largest country if independent, but with only 11.5 million people (that is, less than one person per sq. km), it is very sparsely populated (and for a variety of reasons, that population has been steadily declining since its heyday in the 1960s and 1970s).

To a non-Russian ear, "Arctic Russia" might equal "Siberia," yet the two are imagined quite differently in their native context. From the nineteenth century on, but particularly in the early Soviet period, the Arctic has stood for a place of exploration, the ethnographic encounter with the "other"— the "small people of the North" as the indigenous Arctic populations are termed (see Slezkine 1994)—a place where Soviet and now Russian scientific and military prowess could be tested. But if the Arctic was and continues to be a space of exploration and colonization, Siberia has always been imagined as a place of expulsion and desolation, from the tsarist *katorga* (penal colony) to the Stalinist Gulag (official abbreviation for "Main Administration of Camps," an archipelago of labor camps). Despite constituting about two-thirds of Russia, Siberia in the Russian/Soviet/post-Soviet imagination has always stood for a remote place, far on the outskirts of civilization, a place of exile, forgotten by both time and human memory.

The first recorded voyage to the Russian Arctic was by the Uleb of Novgorod in 1032 and led to the discovery of the Kara Sea. From the eleventh to the sixteenth centuries, Russian coastal dwellers of the White Sea, or *pomors*, gradually explored other parts of the Arctic coastline, going as far as the Ob and Yenisei rivers, and establishing trading posts in Mangazeia. Continuing the search for furs, walrus, and mammoth ivory, the Siberian Cossacks reached the Kolyma River by 1644. The Sea of Okhotsk was discovered in 1639 and the Bering Strait in 1648, with a permanent Russian settlement established in that same period near the present day Anadyr. After Peter I took the throne, Russia began to develop a navy and use it to continue its Arctic exploration. Vitus Bering explored Kamchatka in 1728, while his aides discovered Alaska in 1732. The Great Northern Expedition, which lasted from 1733 to 1743, was one of the largest exploration enterprises in history, organized and led by Bering, Aleksei Chirikov, and a number of other major explorers. They discovered southern Alaska, the Aleutian Islands, and the Commander Islands, and mapped most of the Arctic coastline of Russia (from the White Sea in Europe to the mouth of the Kolyma River in Asia), resulting in 62 large maps and charts of the region. In 1845, Tsar Nicholas I established the Imperial Russian Geographical Society, whose members included explorers, members of the St Petersburg Science Academy, army leaders, and aristocrats. One of its key projects was the creation of the permanent Arctic commission to continue the exploration of the Russian North. The Russian Geographical Society[1] was among the organizers of the first International Polar Year; and stations at the estuary of the River Lena and on the Novaya Zemlya island were created during this time.

The 1840 Russian journal *The Finnish Observer* (*Finskii vestnik*) dedicated an entire issue on the thinking about the north of European Russia, noting that for a long time it presented a "mystery" (*zagadka*) for both central and southern Europe: "some claimed that the North was the cradle of civilization, others populated it with myriads of fantastical nations, and still others extrapolated from it the beginnings of every form of *order* [*vsiakogo poriadka*]." The emphasis on "order" here invokes the myth of Russia's origins, as they were narrated in the twelfth century by the *Primary Chronicle*—ascribed to Nestor, a twelfth-century monk—and which described the origin of the Russian Empire as the moment when a number of northern tribes (both Russian and Finnish), failing to settle their disputes, "invited" Rurik, a Varangian, to "come and rule them."

As Valeria Sobol notes, the ethnicity of the Varangians, from whom the Russian tsars trace their lineage—whether they were seen as Slavs, Romans, Normans, Prussians, Finns, or Swedes—informed the Russian Empire's idea of itself and spoke directly to what Alexander Etkind in his work has termed Russia's "internal colonization": of whether Russia was colonized by its own people—fellow Slavs who imposed order on an unruly population—or by an *other*—be that Roman, Swedish, or German (see Etkind 2011; Sobol 2012). "Internal colonization," as Etkind defines it, speaks to the problem of the state colonizing its own people, but we can take this further and think about the place of Russia-as-Empire in the colonial/postcolonial discourses of the late twentieth century. A contiguous empire, Russia—whether in its imperial, Soviet, or post-Soviet incarnations—is remarkable not only in its history, but also in its geography. Indeed, the Russian Empire is, in a sense, the largest in space *and* the most durable in time of all historical empires: "covering 65 million square kilometer-years for Muskovy/Russia/ Soviet Union versus 45 million for the British Empire" (Etkind 2011: 3). At about the time when the Russian Empire was established, the average radius of a European state was about 160 kilometers. The distance between St Petersburg (established in 1703) and Petropavlosk (1740) is about 9500 kilometers. The Empire was enormous, stretching from Poland and Finland to Alaska, Central Asia, and Manchuria. Its many problems were—and continue to be—partly the result of its size; but throughout the imperial period, tsars and their advisors referred to the vastness of Russia's space as the main reason for its imperial empowerment, centralization, and further expansion (Etkind 2011: 3–4).

But the 1840 *Finnish Observer* was not interested in engaging in the myth of Russia's origins in the North. Instead, it defines the Russian North as comprising, first of all, Finland, and next, the regions of Russia surrounding the cities of Archangel (Arkhangel'sk), Olonets, Volgograd, Perm, Viata, Kostroma, Iaroslavl, Novgorod, and St Petersburg and others. For the *Observer*, the Finns, having originated in Asia, constituted the original population of the Russian North—yet, as the Slavs moved up from the south, the Finns were pushed further and further into the uninhabitable northern reaches of what is now the Russian Empire, leaving behind no traces of their existence. The Finns "cleared the way for the Slavs," (quite literally, by cutting down the forests as they moved north), finally settling in the most inhospitable areas of the North which could not be made habitable to (Slavic) civilization. The *Observer* therefore rejected the notion that Russian tribes might have ever "invited" the Varagians to rule over them, noting that, while it is possible to call upon one's neighbors to help to defend the land against the enemy, it is "against all human nature" to offer yourself to them as slaves. For the *Observer*, Russia's North encompassed the furthest reaches of the Empire, up to the Arctic circle and over to Alaska, the impenetrable tundra of the Arctic regions (both of which serve to protect Russia from invasion), the forests of Siberia (which provide ship-building materials), and Novgorod (the cradle of Russian/Slavic civilization).

As Emma Widdis has noted, by the end of the nineteenth century the official topography of Imperial Russia had a clear shape, with the twin cities of Moscow and St Petersburg anchoring its radial organization, as scientific expeditions into the vast territory gathered information for the center. In cultural terms, however, that territory was still consistently evoked in terms of "unboundedness" and "ungraspability" (*neob"iatnost'*). In the authoritative discourse of an encyclopedia of 1895, for example, it was still possible to write that "vast tracts of the territory of the empire remain technically unmeasured" (Widdis 2003: 6). The exploration of the Arctic has been closely tied in the Russian imaginary to the project of modernity, specifically during the Soviet period, where the forging of the new Soviet state meant an expansion of its ideology into areas previously untouched by imperial Russian concerns. Cinema, in particular, documented the great construction projects that served to connect the vast territories of the Soviet Union into a single unified whole, bringing "civilization"—electricity, railroads, telegraph, newspapers—to the most remote regions.

This chapter provides an overview of the ways in which the Arctic and Siberia have been imagined in the cinema through different historical/political moments of the early Soviet period to the present day. In examining these shifts in representation, my goal is to showcase how the Arctic in the Russian/Soviet imaginary is not static, but has been consistently reconfigured through various historical/ideological paradigms, each intended to erase or reconceive in some way the historical imaginary that came before.

The first Russian footage of the Arctic was filmed by Fyodor Bremer in the regions of the Bering Straight, the Far East, and Kamchatka in 1913–14, the same years that Robert Flaherty first brought a film camera with him on his expeditions. An experienced photographer and cinematographer, Bremer had worked on a number of big-budget productions as well as newsreels, when the studio suggested he take his camera on a polar voyage. In 1913–14, he traveled on the *Kolyma*, which crossed the Polar Circle and became trapped in the Arctic ice. Upon his return, Bremer published accounts of his travels in the film magazine *Pegasis* (*Pegas*) in 1915 and 1916, and a few short films were edited from his footage. Ironically, some of the footage he brought back was damaged, not during the Arctic winter but on the return voyage, which took him south where it was damaged by the heat. Only a few short films survived to be shown in Imperial Russia. Among them was the one-reel *Life of the North* (*Zhizn' severa*, Russia, 1914), showing the interactions of the crew of the *Kolyma* and the indigenous populations (for details on the voyage and the film, see Sarkisova 2015: 222–233).

If, in the Imperial imaginary, the Arctic retained its status as the unknown far reaches of the Russian Empire, for the early Soviet Union it became instead a marker of its expansive reach, to be reconnected with the rest of the Soviet state by means of the railroad, electricity, and the radio. Despite being located in the same geographic space, the Soviet Union was perceived to be a radically different country from Imperial Russia, whose vast expanses were to be rediscovered anew and reconnected with the center. In the early Soviet imaginary, the emphasis was on the production of a dynamic, culturally varied, and rich picture of the young state, and film—in contrast to the museum—could discover "real" ethnographic material by exploring the vast reaches of the USSR. Moreover film, properly used, would be able to escape the fetishized, implicitly colonial vision of ethnic particularity and provide a more "genuine" understanding of the real life of the national republics. This vision would be dynamic and mobile and it would place all peoples on an equal footing, making it "the only place in the world where the many nationalities are in the unique position of total equality" (Widdis: 111–112).

Perhaps the most radical instantiation of this can be seen in Dziga Vertov's *One Sixth of the World* (*Shestaia chast' mira*, 1926). To make the film, Vertov and the Cine-Eye group organized a series of expeditions across Soviet territory, from the Siberian taiga to Dagestan, collecting an enormous amount of documentary and ethnographic material. Vertov's images focused on the specificity and difference of all the cultures represented as part of the Soviet Union, "imagined as independently functioning parts of a greater totality of the state" (Widdis: 110). Vertov's film, however, is not unambiguous about what the North represents for the Soviet Union. Almost half the film is spent on images of the Arctic and, while many of these are what we have come to associate standardly with representations of the North (reindeer herders and small cute children in bulky clothing), we also have images of exploitation and waiting: waiting for the Gostorg ships to arrive, waiting for Soviet power and "civilization," waiting for the moment when our natural resources will no longer be sold off to foreign lands and we can "make machines to make machines."

Overall, the revolutionary quest for knowledge of the wider USSR led to an explosion of ethnographic films during the second half of the 1920s. Cinematography was a powerful tool for visualizing diversity and demonstrating desired developments and achievements. In the Soviet context, the indigenous peoples of the North (and South, and everywhere around the globe) would benefit directly from the new Soviet regime; the landscape they inhabited was imagined as a "complex composite": "a territory rich in material resources and an underdeveloped land; a home to endangered peoples; a vulnerable frontier; and the future venue for an anticipated economic miracle" (Sarkisova: 222).

One of the most interesting examples for our purposes is Vladimir Erofeev's *Beyond the Arctic Circle* (*Za poliarnym krugom*, 1927), which was compiled entirely from archival footage—specifically 10,000 meters of film brought back by Bremer from his expedition on the *Kolyma*. Like fellow documentary filmmaker Esfir Shub, Erofeev had the task of assembling "Imperial" (that is to say, footage shot before the Russian Revolution) material to tell a Soviet story. Erofeev was a great admirer of Robert Flaherty, but he had also spent time in Germany working with filmmaker/explorer Colin Ross, who not only filmed single-handedly in hard to reach places, but also wrote about his experiences/expeditions. *Beyond the Arctic Circle* was Erofeev's first film and the only one in which he relied on pre-existing footage; others that followed (he made 25 documentary films in 13 years) were shot on location in places like Afghanistan and Pamir.

As Oksana Sarkisova (2015) points out, in *Beyond the Arctic Circle* the two editors (Erofeev and Vera Popova) did not draw attention to their use of the archaic 1913 footage, nor did they credit Bremer's camerawork. Instead, they made extensive use of continuity editing, tracking shots, and long panoramas, and added a narrative and intertitles to make the film appeal to the Soviet viewer, but without providing an overt ideological message. As Russian film historian Aleksandr Deriabin notes, Erofeev managed to create a film where Soviet ideology was underplayed in favor of showcasing the original filmic material—to the degree that many critics at the time noted the lack of proper ideological focus, as well as a lack of Vertov-style rapid montage, which made the film appear "old-fashioned" (Deriabin 2001). Indeed in making the film, Erofeev was guided both by the available material, which had been filmed with a largely static camera that remained at eye-level and at a significant distance from the recorded objects, and Flaherty's model of *Nanook of the North* (USA, 1922), which had established certain cinematic conventions for filming the North and its indigenous populations. As a result, Erofeev's film avoided some of the clichés of the standard Soviet ethnographic film, which typically showcased the transformations of the country and the people brought on by Soviet power. Indeed, although Vertov, Shub, and Erofeev are often lumped together into one school of the Soviet "non-played" film, the three had radically different aesthetics. Erofeev was one of the earliest Soviet directors to use panoramic shots consistently, initiating a cinematography of long takes and mobile panoramas. As Widdis puts it, "Erofeev's camera eye was a mobile eye, but it was explicitly the eye of a traveller and explorer," revealing an acute awareness of his own role and that of his team as observer-participants in the world that they filmed (Widdis: 116–117).

While the 1920s produced numerous ethnographic films that explored the far reaches of the new Soviet territory, the 1930s provided new narratives that transformed the Arctic into an integral part of Soviet space to be transformed by Soviet power. Conquering the Arctic formed part of the larger Stalinist projects to remake the world: "to grow Southern plants in the North, irrigate the steppes, etc." (Frank 2010: 115). As Deriabin notes, "in the second half of the 1930s, travelogues disappeared from Soviet screens, and were replaced by cinematic depictions of the courage and heroism of polar explorers, pilots, and "internationalists." Propaganda demanded that the screen reflect only one kind of time—the socialist Golden Age. A "backward peoples" could be shown only from the point of view of the Soviet government's care of them" (Deriabin 1999).

We can see this discursive shift as early as 1935 in the speeches delivered by the members of the newly formed organization, Glavsevmorputi (Glavnoe upravlenie Severnogo morskogo puti/Chief Directorate of the Northern Sea Route), to prepare new cadres of political workers heading to the Arctic. As Otto Schmidt, the head of Glavsevmorputi (and himself, an Arctic explorer, who sailed on the ice-breaker *Sibiriakov*, which along with the *Krasin* and the *Cheliuskin* attempted to navigate the northern sea route) notes, the Soviet Union sent expeditions to the Arctic right away, and already by 1920 they had established the first research center there. But during the 1920s, the exploration and integration of the Arctic into the larger Soviet Union was haphazard and disorganized. It is only with the implementation of the First Five Year Plan that the focus on the Arctic shifted to a "planned economy"—as Schmidt puts it, this was "the beginning of the turn from the first stage of *feeling out* the Arctic, to the present stage of a full frontal assault" (1935: 6). Even though the voyages of the ice-breakers *Krasin* and *Cheliuskin* might on a "formal" level be considered failures—both suffered accidents (indeed, the *Cheliuskin* had to be abandoned altogether)—"politically" these were major triumphs of the USSR, since the rescue "brought the whole nation together and showed what the USSR could do" (1935: 9). Echoing the earlier discourse of the 1920s, but no longer celebrating the difference of the indigenous peoples of the North, Schmidt notes that when encountering local populations the political workers had to provide them with "real culture"—education, boats, and access to machines. It is imperative, he stresses, not to approach the local population in a "museum" way [*muzeino*], as simply an interesting exhibit. Specifically, he notes, "we can capture the shaman on film, but we are going to fight against him" (1935: 20).

Like the American Wild West, the Arctic was a "final frontier" that underscored the relationship of center to periphery in Stalinist discourse: its exploration was part of an all-embracing program to connect every point, every place in the Soviet Union into a unified, homogenized whole by linking even the remotest places directly to the center via radio, electricity, and the like. Susi K. Frank notes specifically that "in all depictions of the Soviet appropriation of the Arctic the figure of the radio operator is of utmost symbolical significance, symbolizing the negation of distances and barriers and the interconnection of distant places" (Frank: 117). Ernst Krenkel, the radio operator of the 1934 *Cheliuskin* expedition and of the 1937 *Papanin* expedition, is a paradigm of this in both Sel'vinskii's epic poem *Cheliuskiana* (published partially in the journals *Novyi mir* (*New World*) and *Oktiabr'* (*October*) in 1937 and 1938) and in Ivan Papanin's "Diary of the SP 1."

One of the major celebrated events of the 1930s was the rescue of the *Cheliuskin* expedition by Soviet pilots. After the first successful journey of the ice-breaker *Sibiriakov* through the Northern Sea Route in 1932, the steamship *Cheliuskin* was to follow to demonstrate that the passage could be made by regular trade ships. Both ships were commanded by Otto Schmidt; the crew of the *Cheliuskin* consisted of 112 people, ten of them women, one of them pregnant, and one child. The ship could not make the passage in time, became stuck, and was crushed by the ice, sinking in February 1934. All of the passengers and crew (except for one man) were able to get off the ship and set up camp on an ice flow where they remained for two months before being rescued in April by Soviet pilots. For this, Stalin created the medal of "Hero of the Soviet Union."

At the same time, however, as Frank notes, Stalinist Arctic discourse was very different from that of the West, which imagined the Arctic as a limiting space: as "an absolute border of the human world, where man is confronted with total loneliness and death as factors of the negation of life, the Arctic resists every form of subjugation" (Frank: 120). Stalinism imagined Arctic exploration instead in colonizing terms: the quest was to establish as many outposts as possible (each reachable via radio waves, where communication hinged on the skill of the radio operator), stocked with film collections, libraries, and other forms of entertainment: "each expedition was conceptualized as a trial for sending as many people as possible and to spend as much time as possible" (Frank: 117). "The conquest of the Arctic" (*zavoevanie Arktiki*) was a key Stalinist ambition, making new marks on the map of Soviet territory, converting the hostile natural world of the Polar Circle into a space penetrated and shaped by Soviet power.

The Vasiliev Brothers (who went on to make that cinematic sensation *Chapaev* in 1934) were chiefly responsible for creating this new image of the North with a feature-length documentary, *Heroic Deed Among the Ice* (*Podvig vo l'dakh*, 1928, also translated as *Exploit on the Ice* and *Ice-Breaker Krasin*), which documented the mission of the ice-breaker *Krasin* sent to rescue the crashed crew of Umberto Nobile's arctic airship *Italia*. The Vasiliev Brothers used the raw footage shot by a cameraman on the *Krasin* to create a narrative of Soviet heroism (their film has not survived, but we do have the directors' notes on the film). These themes of exploration and colonization were picked up by Stalinist cinema, in particular in Vladimir Shneiderov's documentary, *The Two Oceans* (*Dva okeana*, 1933) and Sergei Gerasimov's adventure film, *Seven of the Brave* (*Semero smelykh*, 1936). But even a film like Aleksandr Dovzhenko's *Aerograd* (1935), set in the Russian Far East, is a model for the "long reach" of Soviet power. *Aerograd* is a film about a city of the future, a city that has not yet been built (nor will it have been built by the time the film concludes), about the dream of the Soviet Union's expansion into the Far East, all the way to the Pacific Ocean. The film opens with a single plane flying over the forest and comments on the impenetrability of the Siberian taiga (the taiga is "legible" only to those who live there), and closes with a spectacular sequence of the arrival of Soviet power to this remote land.

But perhaps one of the strangest discourses to emerge about the Arctic during the Stalin period is the discourse of "warmth"—of Soviet power bringing with it not only polar explorers, weather stations, radio operators, Party members, libraries, and cinema—but also climate change. A nice example of this is the book *The Warm Arctic* (*Teplaia Arktika*, 1960) by the journalist Oleg Kudenko, who writes about his travels to the Arctic from 1957 to 1960 (Kudenko 1960). Kudenko opens his book by noting that "like most boys" he had dreamed of the Arctic from an early age—this "land of bravery" ("*krai muzhestva*," which carries with it the connotations of land on the edge of nothingness (*krai*), and also of a specifically masculine brand of bravery (*muzhestvo*)) (Kudenko: 5). He notes that the Soviet government could *not* value the truly limitless possibilities of this gigantic land and that the first assault (*nastuplenie*) on the Arctic began from the first days of the formation of the USSR and that, since 1948, it has continued with renewed energy (Kudenko: 14–15).

Like much of the discourse generated around the Arctic, Kudenko walks a fine line between claiming that the Arctic has been civilized and domesticated, while at the same time underscoring its "strong and brave nature" ("*muzhestvennaia priroda*," 23) that will not be easily conquered: in the beginning of his text, he suggests that "the Arctic obeys mankind, but she is in no hurry to surrender" (Kudenko: 15), and he ends by noting that while "the Arctic is mostly conquered ... the assault on the North continues" (Kudenko: 296). He finishes his narrative by talking about the "warmth" of the Arctic—both in terms of its domestication (it is familiar and homelike, "*domashniaia*," 293), but also, in terms of the climate that over the past few decades, as Soviet climatologists have noted, has become "warmer" and more hospitable. Indeed, he imagines a future in which there will be many cities like Norilsk, and the North will be filled with state farms and greenhouses that will, in his words, "forever alter the climate of the region." He imagines a future of slogans like "Let's Raise the Temperature of the Arctic by 25–35 degrees!" and anticipates that climate change will bring with it gigantic changes of the entire look of the North (Kudenko: 299–300).

But what about Siberia? One of the ironies of the Soviet relationship to the Arctic was that the famous outposts meant to bring civilization to the remote areas were mostly built by convicts expelled by the state. Originally, the vast tundra and taiga of the north were homes to both fishermen and trappers, peasants and Old Believers who had escaped persecution by the Tsar and the Russian Orthodox Church, and the Nenets, Sámi, and Komi peoples. The region was sparsely populated, with one or two people on average per every square kilometer in the Arkhangelsk region, and one person per every four kilometers in the Nenets Autonomous Region (Okrug). As Paul Josephson notes, to this "Soviet rule added two kinds of settlers: exiles and gulag slave laborers. Many of the scientists, explorers, pilots, and captains who explored the Arctic fell into the Stalinist labor camps—the infamous gulag system" (cited in Youngs 2010: 115).

By 1940, there were 53 separate camps and 423 labor colonies in the Gulag. Today's major industrial cities of the Russian Arctic, such as Norilsk, Vorkuta, and Magadan, were originally camps built by prisoners and run by ex-prisoners. There are not many visual records—films or photographs—of the Gulag, but one notable exception is the 1988 documentary, *Solovki Power* (*Vlast' solovetskaia*: the title is a pun on "Soviet power" (*vlast' sovetskaia*)), directed by Marina Goldovskaya. Originally a fifteenth-century monastery founded on the Solovetsky Islands on the White Sea, the place was converted into a labor camp by Vladimir Lenin, and it would ultimately become the model for the Gulag system. In her film, Goldovskaya interviews camp survivors, many of whom were poets, inventors, writers, and historians who underwent "re-education" at the Solovki prison camp. She also interviews some of the guards who prospered under the Stalinist system, and the film includes footage from the

1928 film *Solovki: Solovki Special Purpose Camps* (*Solovki: Solovetskie lageria osobogo naznacheniia*, dir. Andrei Cherkasov), which was shot on location in Solovki labor camp in the late 1920s, ostensibly to prove to the world that the Soviet Union was not using slave labor for its construction projects and that its re-education methods worked. Cherkasov's film was shelved almost immediately after release and was rediscovered in the Soviet archives in the 1980s. Goldovskaya includes footage of a tour of the labor camp system by the famous Soviet writer Maxim Gorky, and documents the daily lives of the prisoners—her *Solovki Power* serves as a kind of commentary on the original Solovki film, including moments when former inmates watch the 1928 propaganda film and comment on its contents.

The Gulag institution was closed by Ministry of Internal Affairs / Ministerstvo vnutrennikh del (MVD) order no. 020 issued on 25 January 1960, but forced labor colonies for political and criminal prisoners continued to exist. In 1972 the camp Perm 36 was converted into the harshest political camp of the country and operated until it was closed in 1988. Until recently, when it was shut down as a museum and closed to visitors, Perm 36 was the only preserved Stalin-era labor camp in the country. Other museums, such as Moscow's State Gulag Museum and Tomsk's interactive NKVD[2] museum, memorialize the Stalinist terror and the Great Purges. But Perm 36 was the only former labor camp that immortalized the lives of political dissidents throughout the entire Soviet era, run by Memorial, a Russian human rights group. In 2012, Putin's crackdown on NGOs threatened to eradicate Memorial and to dismantle Perm 36 as a heritage site, a process that is continuing to this day.

This erasure of the memory of the Gulag is part of the new Russian agenda of accelerated militarization, conquest, and exploitation of the Arctic. In general, as Helge Blakkisrud and Geir Hønneland argue in their 2006 volume, *Tackling Space: Federal Politics and the Russian North*, "after a decade of rapid, ad hoc devolution of power [in the 1990s], the trend has shifted at the turn of the millennium, and Moscow is now emphasizing recentralization and strengthening of the center's political influence" (2006: 15–16). As they stress over and over again, the Russians now have to undo the Soviet project, which emphasized occupation and enlargement (and the desire to turn the Arctic into a garden by conquering nature), but which was entirely too utopian and expansive to do well in a market economy.

As Anindita Banerjee has noted, among Russia's many peripheries, Siberia occupied an especially complicated place in the imaginations of the nation (2012: 23). No one really knew where Russia ended and Siberia began. Unlike the Crimean peninsula or the mountainous Caucasus, Siberia's flat expanses offered no prominent topographical boundaries separating it from the metropolitan centers of St Petersburg and Moscow. As Banerjee argues, "in both geographical and historical terms, Siberia lurked behind the very idea of the nation, playing the role of an uncanny alter ego that perpetually threatened to undermine Russia's efforts to be recognized on the world stage" (Banerjee: 23). She calls Siberia the "negative space in the national imagination"—the lack of physical access to its vast, inhospitable territories, but marked, because of its strategic location and natural resources, by "imperial desire" (Banerjee: 24). It was part of the modernizing project par excellence: on the one hand, a land rich in natural resources and geopolitical significance; on the other, a prehistoric virgin landscape conquered by tracks of modernity. "Rushing past the windows of the train," writes Banerjee, "Siberia was the only place in the world to offer an unlimited view of the deep past preceding human history and the Promethian promises of a technological future" (Banerjee: 33).

The Russian North has always represented both the heart and soul and the *other* of Russia. It has had to be perpetually colonized, and yet it remains the bulk of the country and an unassimilated, uncivilized, unconquered space. That may be one of the reasons why the Arctic/Russian North has also been making an appearance in contemporary cinema, both documentary and feature films, including: Sergei Loznitsa's *Artel* (Russia, 2006), Aleksei Popogrebski's *How I Ended This Summer* (*Kak ia provel etim letom*, Russia, 2010), and Ivan Tverdovsky's *The Island of Communism* (*Ostrov kommunizma*, 2014), among others. These films illustrate how the Arctic region has once again been mobilized for political purposes, and how filmmakers and artists continue to reimagine the Arctic as an alternative space to Russian state power.

NOTES

1. The International Polar Year is a coordinated scientific approach, with observers making coordinated geophysical measurements at several locations during the same year, with 12 expeditions to the Arctic and three to the Antarctic. Russia was one of the twelve nations that participated.
2. The People's Commissariat for Internal Affairs / Narodnyi Komissariat Vnutrennikh Del, abbreviated NKVD, precurser to the KGB.

WORK CITED

Banerjee, Anindita. 2012. *We modern people: Science fiction and the making of Russian modernity*. Middletown: Wesleyan University Press.

Basulto, Dominic. 2015. This Gorgeous Map from 1595 is the key to understanding Russia's current Arctic strategy. https://medium.com/@dominicbasulto/this-gorgeous-map-from-1595-is-the-key-to-understanding-russia-s-current-arctic-strategy-2a5206490202

Blakkisrud, Helge, and Geir Honneland (ed). 2006. *Tackling space: Federal politics and the Russian North*. Lanham: University Press of America.

Deriabin, Aleksandr. 1999. O fil'makh-puteshestviiakh i Aleksandre Litvinove. *Kinovedcheskiie zapiski* 42. Translated into German as: Aleksandr Derjabin: "Aleksandr Litvinov und der sowjetische Expeditionsfilm." In *Die überrumpelte Wirklichkeit. Texte zum sowjetischen Dokumentarfilm der 20er und frühen 30er Jahre*, ed. Hans-Joachim Schlegel. Leipzig: Leipziger Dokwochen GmbH, 2003. Accessed at: http://www.greensalvation.org/old/Russian/Publish/11_rus/11_02.htm

———. 2001. 'Nasha psikhologiia i ikh psikhologiia – sovershenno raznye veshchi.' *Afganistan* Vladimira Erofeeva i sovetskii kulturfil'm dvadtsatykh godov. *Kinovedcheskiie zapiski* 52.

Etkind, Alexander. 2011. *Internal colonization: Russia's imperial experience*. Cambridge: Polity Press.

Frank, Susi K. 2010. City of the sun on ice: The soviet (counter-) discourse of the Arctic in the 1930s. In *Arctic discourses*, ed. Anka Ryall, Johan Schimanski, and Henning Howlid Wærp, 106–131. Newcastle upon Tyne: Cambridge Scholars Publishing.

Kudenko, Oleg. 1960. *Teplaia Arktika*. Moscow: Sovetskaia Rossiia.

Sarkisova, Oksana. 2015. Arctic travelogues: Conquering the soviet North. In *Films on ice: Cinemas of the Arctic*, ed. Scott MacKenzie, and Anna Westerståhl Stenport, 222–234. Edinburgh: Edinburgh University Press.

Schmidt, Otto. 1935. Nashi zadachi po osvoeniiu Arktiki. In *Za osvoenie Arktiki*. Leningrad: Glavsevmorputi.

Slezkine, Yuri. 1994. *Arctic mirrors: Small peoples of the North*. Ithaca: Cornell University Press.

Sobol, Valeria. 2012. Komu ot chuzhikh, a nam ot svoikh: prizvanie variagov v russkoi literature kontsa XVIII veka. In *Tam, vnutri: praktiki vnutrennei kolonizatsii v kul'turnoi istorii Rossii*, ed. Alexader Etkind, Dirk Uffelman, and Ilya Kukulin. Novoe Literaturnoe Obozrenie: Moscow.

Widdis, Emma. 2003. *Visions of a new land*. New Haven: Yale University Press.

Youngs, Tim. 2010. The conquest of the Arctic: The 1937 Soviet expedition. In *Arctic discourses*, ed. Anka Ryall, Johan Schimanski, and Henning Howlid Wærp, 132–150. Newcastle upon Tyne: Cambridge Scholars Publishing.

CHAPTER 11

Invisible Landscapes: Extreme Oil and the Arctic in Experimental Film and Activist Art Practices

Lisa E. Bloom

This chapter focuses on environmental work by activist artists and experimental filmmakers on the polar regions that address through the visual new forms of art, feeling, and sociality that are coming into being in the age of the Anthropocene (arguments here build on those developed in Bloom 1993; Bloom et al. 2008; Bloom and Glasberg 2012). Art and experimental film practitioners are starting to make a significant contribution to the field of Arctic discourses. This chapter analyzes how the artists and filmmakers under consideration develop a unique aesthetic language to explore their concern about the Arctic, the Anthropocene, and how they relate to the concrete realms of the fossil fuel industry, capitalist development, and political notions of territory in the Canadian Tar Sands and the Russian Arctic. They are concerned with the very scope of witnessing, while the extraction of oil changes climates in countries very far afield from where it is produced. These artistic projects also seek to draw connections between the strategies oil companies have used to conceal

L.E. Bloom (✉)
Research Affiliate, Center for the Study of Women, University of California, Los Angeles, USA

© The Author(s) 2017
L.-A. Körber et al. (eds.), *Arctic Environmental Modernities*,
DOI 10.1007/978-3-319-39116-8_11

the world's largest and most unsightly places of resource extraction and processing from view, and the ways that both non-humans and poor and racially minoritized humans have long been disproportionately exposed to the harmful by-products of the oil industry as well as the harshest effects of a warming planet.

INTERCONNECTED AESTHETICS AND POLITICS: URSULA BIEMANN'S *DEEP WEATHER* (2013)

Ursula Biemann is an internationally recognized Swiss artist and video maker whose work has been shown primarily in museums, biennials, university art museums, and galleries throughout the world. The video *Deep Weather* was most recently shown at the 2013 Venice Biennale in the Maldives Pavilion that focused on ecological work from around the world on climate change. Her art practice is strongly research-based, involving fieldwork and video documentation in remote locations. Until recently, she was best known for her work on the gendered dimension of migrant labor, from smuggling on the Spanish–Moroccan border to migrant sex workers in the global context. Her experimental video essays connect a theoretical macro-level with the micro-perspective on political and cultural practices on the ground. Some of her films include *Performing the Border* (1999), *Contained Mobility* (2004), *Black Sea Files* (2005), and *Egyptian Chemistry* (2012). Her video essay format works well for thinking in relational terms and connecting cultures across continents. As such, her work makes us think about embodying a different, more personal, relation to these sites that take into account what Ursula Heise calls a form of "eco-cosmopolitanism" which is "an attempt to envision individuals and groups as part of planetary 'imagined communities' of both human and nonhuman kind" (Heise 2008: 61).

Biemann's practice is simultaneously aesthetic, theoretical, and political. As a video essayist, she is explicitly subjective in her approach (Biemann 2008). Her rejection of a purportedly objective vision remains deliberately incomplete. This resistance to interpretive closure is meant to draw the audience into the urgency of the emotive, poetic, and theoretical aspects of her speech that, in the case of this video essay, is whispered to us in a woman's voice. The first scene of her *Deep Weather*, titled "Carbon Geologies," comprises images photographed from above the Athabasca River that flows north through Alberta, Canada, and into the Arctic

Ocean. These aerial views present a landscape that at first glance seems pristine, but as her footage shifts to the Tar Sands, we see that the entire landscape is irrevocably changed by the overwhelming nightmarish scale of this industrial project (Fig. 11.1).

Industrial terror in these images replaces nineteenth-century notions of the sublime; here the sublime scale is no longer natural but industrial. What now seems so unmanageable is not nature but industry. The video dwells on the unsightly waste of toxic fluids and the dark polluted clouds that hover over the enormous Tar Sands facility to emphasize the environmental damage yet to come that is tied to these vast increasingly obsolescent modern infrastructures.

Whereas her images of the Tar Sands create a stance of distant, cold neutrality in the face of a terrifying spectacle, her text and voice create an intimacy that is a far cry from the transcendent views one associates with these images. What she has to say is whispered to us, as if it is a dirty secret, to suggest pointedly that indigenous populations, both locally near the Tar Sands and elsewhere, are especially harmed by a world changed by hydrocarbons. Even though the camera photographing the Tar Sands pits might be floating above the world, as she tells us her secret we are below and are vulnerable to the enormous environmental and social consequences that will be the legacy of the relentless reach for peak oil energy resources for years to come.

Fig. 11.1 Still from *Deep Weather* (2013). Courtesy Ursula Biemann

In the second section titled "Hydrogeographies," the video unexpectedly shifts (Fig. 11.2) and she takes a planetary perspective to focus on the connection between the relentless search for fossil fuels and the consequences this has more broadly for the far away indigenous Bangladeshi communities located on the Bay of Bengal, one of the coastal regions that has been affected by rising sea levels leaving the shores especially vulnerable to climate change (Amrith 2013). Biemann is interested in the convergent effects of climate change and oil extraction and the negative synergies this produces across vast expanses. The video documents the current struggle of Bangladeshi communities protecting their delta villages from rising sea levels.

In the first part of the video that focuses on Tar Sands the human figure is absent. However, when she takes her camera to Bangladesh she reveals collective human labor as the driving force behind the efforts to shore up and secure a barrier they hope will prevent their communities from being flooded by rising waters (Fig. 11.2).

This section of her digital video shows that the effects are already beyond our imagination as we see footage of the enormous community effort of the Bangladeshis, who only have their own labor to build higher embankments to protect their citizens from such extreme weather events. Land from this perspective according to Biemann "is little more than a constantly fluctuating, mobile mass."

Fig. 11.2   Still from *Deep Weather* (2013). Courtesy Ursula Biemann

Biemann's video understates the dramatic moment of violent storms when the media comes in and often focuses solely on the spectacular apocalyptic aspects of these extreme weather events. Instead, her video is taken before the storm to focus on the collective human preparation of what happens in these flood zones where there is no infrastructure and the majority of people do not have the resources to clean up and rebuild. Her focus on the Bangladeshi's labor is visually and conceptually compellingly designed to prompt the viewer into questioning to what extent cyclones are exceptional and to what extent they are the norm.

In this section, the screen is often split into autonomous parts, deflecting the central perspective of a single frame into multiple perspectives. One example is the image of a young woman who is standing facing us on one part of the split screen that is still, while the other screen depicts an eroding coastline that is now a thin sliver of its former self that is moving in a dizzying way. The female voice-over draws chilling spatial connections with the images as she whispers: "populations along the coastal area drown in their sleep. The signals were muffled and came too late. Fluid lands moved further East and large chunks broke off." This quote draws attention to the potential for devastation when emergency responses fail. The calmness of her whispering voice-over soothes us and belies the volatile issue of climate change in relation to surrounding political and social contexts of reception. This sequence also puts into sharp focus the complex temporality of climate change on a local level. It also provides a particularly haunting example of how the ordinary people of Bangladesh live at the cutting edge of climate collapse.

In his book *Tropic of Chaos: Climate Change and the New Geography of Violence*, Christian Parenti powerfully examines the resulting consequences of anthropogenic climate change and especially extreme weather events already occurring within the belt around the center of the earth between the two tropics (Parenti 2011). He uses the term "damaged societies" in the sense that they, "like damaged people, often respond to new crises in ways that are irrational, short-sighted and self-destructive." However, this is not the case for Bangladeshis in Biemann's video. Rather, what she does show is that the efforts of the Bangledeshis will be woefully inadequate in the long term for preserving their coastlines in the Bay of Bengal, home to nearly half a billion people who are now acutely vulnerable to rising sea level rises (Amrith 2013). Biemann's work gives us an insight into the devastating effects of climate change, especially in the poorer countries, as she whispers to us that "it is no longer to be witnessed here (in Canada) but elsewhere in equatorial zones." For her, the flow of capital in one direction is intrinsically linked to the motion of people in the other. Her work attempts to make sense of this planetary conjuncture within which humanity finds itself today.

Biemann's aesthetic project inevitably comes up against the way in which the dominant media outlets report on poor countries like Bangladesh and her work challenges colonial tropes of the spectacle of so-called "Third World futility and helplessness." She does this by linking Bangladesh to Canada, which works against the reporting of the disaster as just being in the "Third World"; thus she creates a new perception of Tar Sands by showing that its consequences are both far away and local. She also changes our perception of Bangladesh, since many First World audiences do not know the magnitude of the problem there, that substantial parts of the coast line are now submerged into the sea, and increasing numbers of the population will have to live on or near the water.

For artists such as Biemann, nature is no longer a thing apart to be manipulated and exploited at a safe remove. It is now integral to a larger universe of instability, of technological breakdown, social disruption, and suffering that is happening on a planetary scale. Her intention is to make an aesthetic contribution to current discourses about the rapidly evolving dangers of climate change and rising sea levels, and to make intelligible the fact that, if you are living in areas like the Bay of Bengal, the hundred-year flood of yesterday is now the monthly event of today.

Landscapes for Biemann are not interpreted as natural phenomena or as venues for events; rather they are important in her efforts to write counter-geographies into these scientific planetary scripts and to elucidate the link between the peoples and histories of Canada and Bangladesh. By comparing Bangladesh with Northern Canada, Biemann also makes us consider the difference in the carbon footprint between richer and poorer nations, and makes us question whether it is really fair to speak of the climate change crisis as a common "human" concern. The question of differences leads us back to power, to the politics of locations and the necessity of an ethical-political theory of subjectivity and naming, asking us to question who exactly is the "we of this pan-humanity bond in fear of a common threat?"

THE MELODRAMA OF HYPER-REALITY: BRENDA LONGFELLOW'S *DEAD DUCKS* (2012)

Brenda Longfellow is a well-known Canadian filmmaker, writer, and academic, whose films primarily circulate at international film festivals and, in some cases, are shown on Canadian television. She teaches in the Department of Cinema and Media Arts at York University in Toronto, and her writings and films are taught in university film and communication departments. She is co-editor of the anthology *Gendering the Nation: Canadian Women's Cinema* (Armatage et al. 1999). As a filmmaker she, too, focuses on representing the Tar Sands of her country. Her 2012 film *Dead Ducks* is the second piece in a trilogy of art projects she has made about proliferating oil mega-projects.[1] The other films that are part of her trilogy include *Carpe Diem* (2010) and the interactive documentary *OFFSHORE* (2014). *Dead Ducks* uses opera and animation as a way of satirically focusing on how to represent the challenge and difficulty of dealing with the ecological devastation that is happening at Tar Sands without deploying a solely human point of view. The story of her documentary is based on a real event about 1606 ducks that travel from Louisiana to Alberta only to die a horrible death by drowning in the oily muck of a Tar Sands settling pond. Like Biemann's whispering voice in her film, the "voice of god" narration common in more conventional documentaries is replaced in Longfellow's feminist film by the point of view of the birds as well as those humans that comment on them: a female veterinarian, an environmentalist, and an indigenous worker who has divided loyalties between his job as an engineer at Tar Sands and his community and family who have relied on the ducks in the past for food. *Dead Ducks* is both a serious documentary and a critical piece on environmental hazards connected to Tar Sands. Like Biemann's video it experiments with an aesthetics of intimacy, though in this case it uses animation to represent the ducks that travel from place to place before they end up dying in the ponds. Longfellow creates an embodied sensual experience of the birds. The viewer is brought so close to the animated birds that we can almost see their faces as they soar hundreds and thousands of feet above the earth, and the animated filmmaking technique allows us to feel as if we were flying right along with them (Fig. 11.3).

With her use of music, she experiments with both vision and sound to create new forms of film language to depict a post-natural condition that is empathetic to the birds, but at the same time critical of the way we anthro-

Fig. 11.3   Still from *Dead Ducks* (2012). Courtesy Brenda Longfellow

pomorphize and over-identify with the non-human. She also uses X-ray, remote sensing, and other visualizing technologies that are used to track and monitor the birds' migration movements that have been impacted on by changes in the weather and more recently scarcities in water and food. The images of the birds are both "real" and artificial and are meant deliberately to reference popular films like *Winged Migration* (2001) where CGI birds were inserted into a documentary film to generate viewers' sympathy with an animal protagonist. Partly, she does this to get us close enough to the birds to capture both the mundane and resplendent aspects of their lives, but also to have the audience empathize with them when they are caught unaware of the toxic waste in areas where they might migrate to for food and water now that their migration patterns are changing because of climate-change induced drought. Their hyper-real and color-filled landscape is juxtaposed with the dismal "real" industrial sublime landscape of Tar Sands represented in the film in black and white to signify that the ducks are unable to differentiate between clean or polluted water. In a certain way *Dead Ducks* is an imaginative response by the filmmaker to attempt to make sense of the intense public outcry and the response by the oil company over the plight of the ducks. What is significant about the duck incident for Longfellow is how visual documentation by Todd Powell (a senior wildlife biologist with the Alberta Government) featuring ducks dying in Tar Sands pond went viral. These images helped to galvanize international public opinion against the Tar Sands project at a moment when senior ministers from the government were lobbying American senators to adopt the view that the Tar Sands pipelines represented the solution to American energy security (Longfellow 2013, 2018). *Dead Ducks* proposes that, despite the publicity given to the birds, multinational oil corporations are quite adept at managing ongoing environmental disasters as mere public relations crisis, and of dissipating an ecological crisis and turning the tragedy into something that can be managed and contained. The male spokespersons for the oil company in Longfellow's work deliberately use traditional gender roles through dress and speech to reassure us that they will take all responsibility and fix the situation. As we will see, this will be the target of the activist group called The Yes Men. Both Longfellow and The Yes Men remind us how successful the public relations arms of these companies are and the challenges for activists of finding ways to represent environmental crises that aren't contained by the rehearsed performance of an earnest public apology by these companies that months later is easily forgotten by the public. In this case, while the reputation of Syncrude, the oil company connected to the Tar Sands project, was hurt, in the end the company was fined just $1 million, and what was ultimately forgotten was that the real and ongoing disaster of unconstrained oil extraction was allowed to flourish. Longfellow, in a nod to Biemann's video, seems to understand how the ducks enter into the more ordinary realm of a melodrama of helplessness and as an object of pity in this political context when she pointedly asks: "How was it that the plight of these ducks could evoke such emotional response when the plight of millions of Bangladeshis, coastal inhabitants, Inuit and sub-Saharans left most Canadians indifferent?" (Longfellow 2018).

Absurd Impersonations: *The Yes Men: "But It's Not That Polar Bear Thing"* (2013)

Biemann and Longfellow aspire to make post-natural landscapes for our time and provide an insight into gender, human labor, and human and non-human lives at the sites that are impacted upon. In contrast, The Yes Men's activist art performance pieces are often more specific and targeted in their unveiling of the hiddenness and secrecy at play in the media's representation of industrial mega-projects. The Yes Men's activist work is widely known within contemporary art circles and is shown at both international art exhibitions and more widely through activist venues. Their work is also widely shown at universities in Media, Art, and Communication departments. The two leading members of The Yes Men who live in the USA are Jacques Servin, an author of experimental fiction, and Igor Vamos, an associate professor of Media Arts at Rennselaer Polytechnic Institute, New York. The two films that they have produced include *The Yes Men* (2003) and *The Yes Men Fix the World* (2009). In these films and in their performance work they often impersonate corporate or governmental figures that they believe are acting in deliberately dishonest ways to further the agendas of the companies or governmental organizations where they work. The Yes Men often also create and maintain fake websites to play with and subvert the public image of the various companies that they target to undo the very careful way these companies script their visual representation. In 2012, The Yes Men collaborated with Greenpeace and members of the Occupy Seattle movement to focus on issues dealing specifically with Shell's oil drilling activities in the Arctic Circle in their recent website (www.arcticready.com). More recently, in their video *The Yes Men: "But It's Not that Polar Bear Thing"* (2013) they staged an elaborate spectacle and a fake PR campaign for Gazprom, the Russian Gas company which has partnered with Shell. Their performance involved taking over a barge in Amsterdam with an apparently drugged polar bear, a Russian child superstar, and a marching band that moved through the city's canals to the zoo where the artists, dressed as corporate executives, made the gift of the drugged polar bear. The Yes Men's work here is focused on the distasteful attempts of oil companies to fool the public and to challenge the monological narration of oil. The tightly scripted PR campaigns of oil companies like Gazprom and Shell convince the world that they are not responsible for the collapse of our ecosystems, that they can be trusted to drill safely in the Arctic without ruining it and reducing living standards for humans and non-humans and that oil drilling does not speed up the melting of ice by climate change. Since the Arctic is already warming at more than double the rate of the rest of the planet, these insights resonate in a global context and are one of the ways that drilling for oil has evolved into an issue of justice on a planetary scale, since the Arctic's future is so crucial to a livable planet for humans and non-humans. This is also why Subhankar Banerjee's important edited book *Arctic Voices* underscores the urgency of this issue in the Arctic with its subheading "Resistance at the Tipping Point." The Yes Men's performance pointedly mimics Gazprom's own cynical PR campaign by trotting out the gift of a drugged "polar bear," the global icon of climate change, as a quick way to demonstrate their apparent concern with climate change. The gift of the polar bear contributes to duping the audience into believing that this is a genuine effort on the part of Gazprom and Shell to show that they empathize with the plight of polar bears and that they care about global warming.

CONCLUSION

All three of these art projects question what it means to live in an age of secrecy and the ways in which oil companies have constructed an invisible power structure in their media campaigns that have successfully managed to conceal the world's largest and most unsightly sites of resource extraction and processing from view. They are also concerned with the consequences of what this industry is doing to the humans and the non-humans most impacted upon by climate change. Furthermore, the case of The Yes Men's activist art performance places emphasis on the gullibility of the public to the green-washing of the oil company's public relations and the need for a visual depth and complexity to understanding the relation between climate change and the world of oil.

Today, in an age that mostly celebrates instant spectacles, the relatively slow and open-ended aspects of climate change represent obstacles that can hinder efforts to mobilize citizens to appreciate the urgency of the situation and to think differently about the long-term consequences. The complex temporality of climate change as a future projection that lacks any finite end makes this especially difficult for artists to represent effectively.

The significance of their contributions is their attempt to jolt dramatically the standard perception of when and where climate change is supposed to happen as our addiction to burning fossil fuels activates profound changes in the planetary ecology. That is why artists like Biemann and others are focusing on what is happening to indigenous peoples in areas far from metropolitan centers where climate change otherwise appears invisible as it is often seen as one more disaster added to the ongoing intractable problems of poverty, colonialism, and underdevelopment. Thus one of the tasks of this chapter and of my ongoing research is to present more complex images of global warming and environmentalism that are neither apocalyptic nor sentimental by avoiding the typical iconography of crashing glaciers and melancholic polar bears that dominate the visual cultures of climate change discourse or the scenarios depicted in disaster films such as *The Day After Tomorrow* (2004) among others. My hope is that new perspectives on visualizing climate change as represented by these artists will help us conceptualize human fears about the geophysical "end of the earth" catalyzed by our entry into the Anthropocene. By asking other questions, such as who the indigenous of the Anthropocene are, and how we might represent the multi-scale, multi-temporal, and mutual connectedness of this epoch, these artists tell different stories to make us aware of how our entire way of thinking and being is now undergoing a melting transformation. Such work is especially necessary if we are now to imagine the "we" that humans are supposed to feel a part of in taking responsibility for the Anthropocene.

Note

1. Longfellow's *Dead Ducks* is available through open access on Vimeo: https://vimeo.com/37867483.

Work Cited

Amrith, Sunil S. 2013. The Bay of Bengal, in peril from climate change. *New York Times*, 13 October 2013. http://www.nytimes.com/2013/10/14/opinion/the-bay-of-bengal-in-peril-from-climate-change.html?pagewanted=all&_r=0

Armatage, Kay, Kass Banning, Brenda Longfellow, and Janine Marchessault (ed). 1999. *Gendering the nation: Canadian women's cinema*. Toronto: University of Toronto Press.

Banerjee, Subhankar (ed). 2013. *Arctic voices: Resistance at the tipping point*. New York: Seven Stories Press.

Biemann, Ursula. 2008. *Mission reports: Artistic practice in the field, video works 1998–2008*. Umeå: Bildmuseet, Umeå University.

Bloom, Lisa. 1993. *Gender on ice: American ideologies of polar expeditions*. Minneapolis: University of Minnesota Press.

———. Forthcoming, 2018. *Polar art in the anthropocene: gender, race, and climate change*. Durham: Duke University Press.

Bloom, Lisa, and Elena Glasberg. 2012. Disappearing ice and missing data: Visual culture of the polar regions and global warming. In *Far fields: Digital culture, climate change, and the poles*, ed. Andrea Polli, and Jane Marsching. Bristol: Intellect Press.

Bloom, Lisa, Elena Glasberg, and Laura Kay. 2008. Introduction to special issue, gender on ice: Feminist approaches to the Arctic and Antarctic. *The Scholar and the Feminist* 7. Available at http://www.barnard.edu/sfonline/ice/intro_01. htm

Heise, Ursula. 2008. *Sense of place and sense of planet: The environmental imagination of the global.* Oxford: Oxford University Press.

LeMenager, Stephanie. 2014. *Living oil: petroleum culture in the American century.* New York: Oxford University Press.

Longfellow, Brenda. 2013. *OFFSHORE: Extreme oil and the dissapearing future. Public: Art/Culture/Ideas* 48: 95–104.

Longfellow, Brenda. 2018. Extreme oil and the perils of cinematic practice. In *Petrocultures: Oil, Energy, Culture*, ed. Sheena Wilson, Adam Carlson and Imre Szeman. Montreal: McGill-Queen's University Press.

Parenti, Christian. 2011. *Tropic of chaos: Climate change and the new geography of violence.* New York: Nation Books.

CHAPTER 12

Icelandic Futures: Arctic Dreams and Geographies of Crisis

Ann-Sofie Nielsen Gremaud

Icelandic society is still marked by the economic collapse of 2008, and the subsequent crisis has led to heated discussions about the future of the country. In the aftermath of the collapse, several official initiatives had been launched to seek to link the country more closely with the Arctic—a space that attracts international attention associated with ambitions to extract and control valuable resources, while also being a location of environmental concern as effects of climate change are increasingly evident in the region. The Arctic has effectively become a space for the articulation of crisis management in the political rhetoric of the Icelandic Government (2013–) and in the rhetoric of former President Ólafur Ragnar Grímsson (1996–2016). Most notably, Grímsson has outlined the opportunities for an Icelandic Arctic future in recent speeches: "After being isolated for centuries, and held tight in the shackles of the cold war during its first decades as an independent republic, Iceland is now a much-sought-after partner in the growing collaboration over the New North. ... It is a blessing for a small nation that now stands on the threshold of a new era, after coming back onto its feet following the collapse of its banks, to have the opportunity to embark on such a journey" (2014). In his speech at the celebration of the West Nordic Council's 30-year jubilee in

A.-S.N. Gremaud (✉)
University of Copenhagen, Copenhagen, Denmark

© The Author(s) 2017
L.-A. Körber et al. (eds.), *Arctic Environmental Modernities*,
DOI 10.1007/978-3-319-39116-8_12

2015, President Grímsson highlighted Iceland's privileged position in the region, while emphasizing environmental history and politics as a matter of heritage in the Arctic: "inculcated in us by the experience and wisdom of our ancestors" (2015).

In Iceland as elsewhere, the notion of "The Arctic" serves as a global carrier of discourses of climate change and geopolitical positioning and as a prominent scene of identity negotiations. The development in Iceland sheds light on the relativity of the Arctic as a region and a space that is defined by projections of visions, hopes, and fears. The existing stock of associations connected with the high North is influencing the image being promoted by official Icelandic sources and in turn a number of unofficial reactions have created a field where this image and its implications are negotiated. Through analyses of visual art and public visions for the future, this chapter examines Arctic environments as a frame for imagined futures. To this end, I pose the following questions: What landscapes and thus what (Arctic) ecologies are imagined in visual and textual discourses in contemporary Iceland? Are the present and future ecologies imagined as continuations of or breaks with the ways of the past? What geographical levels (local, regional, global, and planetary) are in focus? And, what pitfalls do the artworks point to in current strategies for future Arctic involvement?

A central point of disagreement in Iceland concerns the management of natural resources as the foundation for a future society. Art can be a laboratory of environmental theory because of its ability to present alternatives to reigning ideological patterns through effective intervention or semiotic analysis. Political and critical art represents a creative sphere where new perspectives are opened, which to some extent replaces a critical Icelandic public sphere for political dialogue and critical journalism that has had hard conditions in Icelandic society. Criticism of the agenda of official economic and environmental policies can be found in artworks from the decade leading up to the collapse and in the years following it. Fundamentally different interests and visions embedded in the conflict about attitudes to nature shed light on the contradictions in official narratives about the environment that could spell future challenges to engagement in the Arctic region.

The aftermath of what was simply known in Iceland as "the collapse" (in Icelandic: *Hrunið*) still affects decision making, discourses about the future, and views on history. The ongoing dispute about the future role of nature and natural resources is intertwined with interpretations of the economic, cultural, and environmental crisis, and the uncertainty after the shock of the collapse is a driving force behind several contemporary agendas. Both in the years leading up to the economic collapse in 2008 and in the following years, environmental policies have been widely addressed in scholarly studies, artistic commentary, and political debate. Icelandic political leaders are positioning the country within the Arctic region and building visions of the future within this geography, while many artists have focused their attention on unresolved planetary and national issues. While Icelandic environmental politics seem to be dominated by a value system based on material accumulation and consumption that promotes the ideal of a utilitarian approach to natural resources, many artworks are eco-critical and promote an ecology in line with the environmental philosophy of Piers Stephens, who promotes an anthropocentrism "which thinks of the human agent as a many-sided agent rather than solely a consumer" (2000, 287) and who emphasizes the environmental responsibility of the Western world.

Art has the ability to open spaces of subversive potential and, unlike most political discourse, to engage in sustained critique. In a context where people are largely viewed as consumers, it is not surprising that some artists engage in a critique of this condition and attempt to imagine alternative relationships between society and nature. Both in a national and a planetary framework, Icelanders are faced with questions about sustainability and solutions for the future. Icelandic artist Ásmundur Ásmundsson's *Into the Firmament* (2005) addresses the interconnectedness of these geographical levels. The installation, consisting of oil barrels and cement forming a tall pyramid, was briefly exhibited in the public space at the peak of economic optimism and investment in Iceland. Both the construction and the title contain a clear reference to the mythical Tower of Babel. The oil barrels, overflowing with cement, suggest that a limit is being reached. His work *Hole* from 2009, a remake of a 2006 performance in Viðey, is another monument of infamy. In the political context of the recent collapse, children from Reykjavik were invited to dig a hole in the ground that was then filled with cement. The subsequent cast sculpture of concrete and steel, measuring 2 × 2 × 2 meters, was to represent the economical abyss that future generations were to inherit and somehow work their way out of.

The official material and artworks addressed in this chapter refer to different geographical levels: from the local environment or the national level to the planetary and global level (with a focus on the one hand on environmental questions and on the other hand on globalization, the market economy, and geopolitical networks), reflecting variations in emphasis on different agendas, hopes, and fears. The artworks directly and indirectly address potential environmental risks inherent in the official strategies, as the High North is a scene for ongoing negotiations of the imagined geography of the Arctic as well as renegotiations of the role of natural resources (oil, fish, hydropower). These processes of negotiation and renegotiation also pertain to ideas about national identity in former Danish colonies, which are young nations still in the process of negotiating fundamental values in a global framework. An Arctic strategy from the Faroese Office of Foreign Service was presented in early 2013. The very title, "The Faroe Islands: A Nation in the Arctic," holds a clear message. A key issue in Icelandic foreign policy, as presented in the government's policy declaration of 2013, is to position the country as a future "leading power" (*leiðandi afl*) in the Arctic region (Declaration of Policy, Utanríkismál, 11).

The legacy of the Icelandic independence movement continues to influence both domestic and international policies. Drawing the line between the inside and outside of the national collective included a strategic emphasis on the close connection between a united people and the land. Some have suggested that the distribution of control over the resources is inherited from earlier structures—from the time when Iceland was a part of the Danish Realm and even from the age of the chieftains, when power was held by a small elite; a situation that led to persistent administrative nepotism (see Erlingsdóttir 2009; Hafstein 2011). Historian Guðmundur Hálfdanarson has pointed to the remarkable durability of the myth of the unified nation (2000). Celebrating the Parliament's agreement on Iceland's Arctic policy then president Ó. R. Grímsson ended his 2014 New Year address with a reference to post-apartheid South Africa: "we in Iceland also have our store of accumulated wisdom, the experience of our history to serve us as we journey forward, from the conflict of recent years towards lasting cohesion and solidarity" (Grímsson 2014). The argument of the speech discreetly links the unified nation as a means to overcome colonial injustice with the state of crisis after 2008. By framing the turmoil and internal division through a discourse of reconciliation the anger and criticism of the Icelandic individuals who lost money as well as trust in their political system became a generalized and abstract problem of the national collective. As the origin of the problems as well as their solution melts into an intangible collective sphere so does the question of responsibility.

In this way the new focus on the Arctic region—including closer ties with Greenland—helps place a proactive, united Iceland at the center of the world's attention, where it has recovered from hardship and an "apartheid-like" inner division and eventually overcome the internal distrust which, according to the former president, is what hinders growth. One might polemically ask whether this view on history, that supports Halfdanarson's thesis of the myth of a unified nation, might halt critical scrutiny of the collapse.

THE HIGH NORTH AND THE IMPORTANCE OF BEING NORDIC

Discourses about the High North or the Arctic have been given many labels, including "arcticality," "borealism," and "norientalism". Historian Sumarliði Ísleifsson has traced long-lived stereotypical ideas of the North (2011) of which I find the notions of "the utopian North," "the original North," and "the wealthy North" of particular relevance in an analysis of the discourses of Arctic Iceland. Ísleifsson outlines a general development over recent centuries that defines the countries north of Scandinavia as the "Far North" or the "High North." From the 1700s, the High North became increasingly connected to what has been perceived as the civilized center, "if at the edges of it" (Ísleifsson 2011, 15). Discursive constructions during the Enlightenment and the period of nineteenth-century national romanticism had, however, cemented a division between Iceland and the Faroe Islands as the Germanic Far North versus Greenland and the Sámi as the indigenous peoples (16): a logic that is being challenged by the current Arctic region building.

Romantic influence on the Icelandic image has linked it with the utopian North (living in balance with nature) and the original North (a living past and a shrine of Nordic cultural heritage). This influence is reflected in the poetry and texts of the canonical figures of national romanticism such as N. F. S. Grundtvig, Adam Oehlenschläger, and the politician Orla Lehmann, who described Iceland as "a living antiquity, a talking image of the life of the past" (524).

The idealization of cultural purity that is linked with the ideas of the original and utopian North has led to this persistent tradition of Iceland being associated with the past—specifically with Old Norse heritage (see Gremaud 2014a). In the early twentieth century, Iceland experienced rapid urbanization, which influenced the country's branding strategies, as evident in its representation at the New York World Fair in 1939, "The World of Tomorrow," where focus was explicitly on the future. In his description of the third stereotypical idea (2011)—that of the wealthy North—Ísleifsson refers to the writings of chronicler Adam of Bremen (c.1040–81), who described a land in the North "where gold and gems were in abundance and where the inhabitants possessed only a rudimentary understanding of this wealth" (Ísleifsson 2011, 13). Today this idea of the North as a place rich in resources is reflected in the optimism of the global race for the resources of the Arctic underground. The wealthy North may be said to include a notion associated with the High North on the one side (that of resource extraction) and a notion associated with Scandinavia on the other (that of the privileged Nordic countries) (*The Economist* 2013). Current official statements and branding strategies form a strategic oscillation between associations with stereotypical and romanticized ideas of the original and utopian North on the one hand and Arctic Iceland in the making as a materialization of the wealthy North on the other. The representations and ideas of the North are in a reciprocal relationship with ideas from nation-building processes and thus with political ideas and the power play of jockeying for a favorable position in the geopolitical system. The political and cultural history of Iceland and the immediate region surrounding it has laid the foundation for two primary temporal axes at play in dominant national narratives: a vertical ethnocentric axis of originality and a horizontal axis related to progress. These axes and stereotypical notions of the North affect current discourses about natural resources, political dispositions relating to visions for Iceland's role in the Arctic.

Since the middle of the twentieth century Icelandic society experienced great economic change, moving from a position as a developing country and Danish dependency to a position around the beginning of the twenty-first century when it gained global attention for its aggressive investments before the devastating economic crash of 2008. The narratives reflected in today's policies have developed in a culture characterized by crypto-colonial features influenced by the strong currents of nationalism, Eurocentrism, and industrialization; thus the symbolic value of nature has been greatly influenced by the nation-building process (see Gremaud 2014a). Only in very recent times have views on natural resources become increasingly influenced by theories of the Anthropocene, spurring inevitable conflict about priorities and responsibility in the same way that, for example, a number of artists have begun to criticize the dominant capitalist and neo-liberal visions that are a further development of the narratives of the nation-building process.

The (new) explicit identification with the High North evokes historical issues of power related to being on "the right side" of an imperialist duality. In a historical perspective, the divide between a perceived Germanic North and the Inuit culture has influenced the regional relations between Iceland and its Arctic neighbor Greenland. According to Ísleifsson (2009) these hierarchical categories as a means of tying parts of the High North closer to Europe while turning away from others also influenced Icelandic self-representation—especially from the twentieth century onward: "the Icelanders took this image [the Germanic High North] to heart, an image partly characterized by ideas of superiority and racism; it became the self-image of the Icelanders for decades before and after independence in 1944" (Ísleifsson 2009, 154). The ambivalence and insecurity associated with a crypto-colonial position (see Gremaud 2014a) on the perceived borders of the civilized world resulted in a distancing from those associated with the indigenous far North famously expressed in the dispute over the inclusion of Iceland in a colonial exhibition in Copenhagen alongside "Negros and Eskimos" in 1905 (see Jóhannsson 2003). Furthermore, Hálfdanarson has shown that a key argument in the discussions in 1911 about founding a university in Iceland was the ambition of consolidating the country's status as a civilized society, equal to other nations (it was not, however, declared a sovereign republic until 1944). Iceland eventually left the Society of Atlantic Islands, referred to as "the Society of Barbarians" (Hálfdanarson 2011, 301) in the parliamentary debate at the time, thus seemingly escaping a negative regional identification. The current policy, which aims to position the country as a regional leader, reflects the perceived potential of the Arctic and could mark a step away from strategies of self-exotification (see Schram 2009); this includes a tongue-in-cheek strategic essentialism that has been promoted in recent commercial visual culture. Thus, for Iceland, the Arctic has become a new scene for geopolitical positioning away from the country's stance as an outsider in the European context. Political scholar Valur Ingimundarson has stated, "indeed, Arctic power games are about identity politics—about exclusion and inclusion—whereby states and organizations are classified on the basis of power and legitimacy as those on the inside and those on the outside" (2011, 189). This points to a key element in the Icelandic ambition to move further away from a position of geopolitical insignificance and toward equality and recognition.

(Arctic) Iceland: Branding Purity

As outlined above, two intertwined political fields make up important contexts for current environmental and geopolitical discourses in Iceland: the making of the Arctic region and the aftermath of the economic collapse. In the efforts to tackle the national crisis one can observe the making of Arctic Iceland. Old notions of the unspoiled Northern wilderness have been reintroduced with concepts of Iceland used in branding strategies and in political and corporate visions of the future relationship between nature and society. By virtue of its associations with the original North, Iceland is being lined up within the Arctic as a North par excellence, or a super-North. The government that took office in 2013 proposed a withdrawal of the country's application for membership of the European Union (EU), reflecting a long-standing euro-skepticism. As of 2014 negotiations with the EU have been paused, and the Arctic region has become a primary focus area in foreign policy.

Iceland's official Arctic strategy responds to both political and environmental changes, and this game of geopolitical positioning includes messages with multiple layers of intertextual references. An introductory statement in a 2009 report from the Ministry for Foreign Affairs positions Iceland as "the only sovereign nation to lie entirely within the Arctic" (*Arctic Report/Iceland in the Arctic*, 7). A few years earlier, the Ministry published a report from the conference entitled "Breaking the Ice," which included an opening sentence signaling Iceland's interest in mobilizing the following concept: "we who live in the Arctic region" (Icelandic Government 2007, 1). On the cover there is a drawing of a Viking vessel carrying a band of Vikings with raised weapons, an image that evokes the Icelandic national narratives of *útrás* and *landnám* (expansion and conquest) by depicting Icelanders as active agents in the oceanic space that constitutes the Arctic region—the much anticipated expanding shipping routes. The then Prime Minister (2013–16) and leader of the Progressive Party, Sigmundur Davíð Gunnlaugsson, recently further sparked this optimistic vision for Iceland's future position as part of a privileged Arctic region. With a reference to Laurence C. Smith's book *The World in 2050: Four Forces Shaping Civilization's Northern Future* (2011) Gunnlaugsson focused on making the most of climate change: "and Iceland was one of the eight countries of the future. It is highlighted that many opportunities are obviously opening up in the Arctic for shipping routes for oil and gas and other raw materials, and not least for food production" (Progressive Party Website, author's translation).

Photographer Ragnar Axelsson presents another side of this vision in his photo book *Last Days of the Arctic* (Axelsson and Nuttall 2010). In the political visions for Arctic Iceland, environmental protection is not presented as a contrast to resource extraction; rather, his photographs show the Arctic as a place where a price is being paid for anthropogenic climate change. With a focus on Inuit culture, human beings are represented as (small) parts of a vanishing ecosystem, and the Arctic becomes a scene for the ways of the past. In the introduction to *Last Days of the Arctic*, anthropologist Marc Nuttall describes how indigenous temporality is portrayed as time standing still, and the Arctic as a place where one can rediscover one's own footprints from decades ago (14). The black-and-white images support a notion of timelessness and association with the past, while awareness of the transformations following the melting of sea and glacial ice makes these photos into a statement about the Arctic being on the tipping point of irreversible change. Axelsson's low camera perspective and use of contrast and light/shadow effects make the mountains and hunters appear monumental, lending a mythological air to some of the images. Framing and motifs showcase the Arctic as a place where humans live at the mercy of nature. In Axelsson's images, ice, the symbol of purity in several Icelandic strategies, becomes both the scenic frame as well as the disappearing subject in focus.

The performance group, The Icelandic Love Corporation, has also examined the role of the high North as a dreamscape that merges dystopian and utopian features. In their artwork *Dynasty* (2007) (video and photographs), that shows a performance taking place near the hydro-electric power station Vatnsfell, the issue of climate change provides a regional macro-context for questions about the conditions for human life in the future. Here, the High North becomes a place for imagining future engagement with nature in a post-climate-change setting where material luxury becomes redundant, symbolized by the fur-clad women's act of burying their jewelry and cell phones. This reflects a vision of a future where the High North has become an exclusive and refreshingly cool location in a world where snow and ice is rapidly vanishing. Two temporalities collide in this artwork, which comes to support its critical potential: the slow tempo of the film as it lingers on the women's engagement with the mundane tasks of hunting, fishing, and guitar playing in the snow-covered landscape on the one hand and the urgency of climate change that makes this location a unique destination on the other.

Fig. 12.1 The Icelandic Love Corporation: *Dynasty* (2007). 9 photographs, ach 70 x 100 cm. Edition 3. Courtesy The Icelandic Love Corporation.

Official statements in the branding strategies pursued by state-owned companies and the tourism industry link Iceland's energy sector and the general Icelandic nation brand with purity. Statements and imagery reflect an anthropocentric value system inherent in these fields dominated by a concern for the country's image and a framing of nature as an energy resource, which was also reflected in the first political program of the government that took office in 2013: "nature is one of the country's main resources," and "pure renewable energy" will benefit both export and "the strong image of the country" (Declaration of Policy, Umhverfismál).

On the website of Iceland's largest energy provider, Orkuveita Reykjavíkur, purity is a key implied value. The website features a short film titled *Pure Nature* (*Hrein náttúra*) about the company's environmental policy. This image is also reflected by "Iceland Naturally," the official branding portal for food producers and the tourism industry, launched in the United States in 1999 and in Europe in 2006, where it is stated that ice "represents the source of our pure water and symbolizes the purity of all Icelandic products. Indeed, nature is our brand and Iceland is dedicated to preserving this natural wealth through responsible conservation" (Iceland Naturally 2013). With a website characterized by a strong focus on purity, "Iceland Naturally" conveys a message with a different focus than that of the new governmental policy on environmental issues: "the government will, as far as possible promote the utilization of potential oil and gas deposits to begin as soon as possible, if they are found in extractable quantities" (Declaration of Policy, Olía og gas). Both statements, however, support an optimistic discourse about extracting natural resources in a way that upholds an image of a pure industry (see Gremaud 2014b). Together with the branding statements and videos promoted by Orkuveita Reykjavíkur, "Iceland Naturally" is arguing for a trinity of purity, energy production, and the national image that is directly and indirectly supported by other discursive fields such as tourism, branding, design, governmental policies as well as external representations of Iceland. The 2008 report also suggested initiatives to strengthen internal consensus: "it has to be a collective task of the nation to protect the image and bring forward the right message" (Branding Report/Iceland's Image, 14). Geographer Edward Huijbens points out that this effort involved artists producing positive stories in a manner that would position the artworks as illustrations in a marketing strategy (2011, 564). In his assessment of the report, he concludes that "power and purity are suffusing landscape myths, transposed onto the inhabitants" (570). This is a statement that may find support in Icelandic visual culture where untouched landscapes have been a favored motif since the onset of national romanticism in the nineteenth century. Thus notions of pristine, pure, original nature are linked with nation-building and a naturalized part of the nation brand. However, some artworks question the narrative of the harmless trinity by showing hydroelectric and aluminum industries as examples of primary destroyers of wild nature in Iceland. Artist and mountain guide Ósk Vilhjálmsdóttir has addressed the consequences of resource extraction in the energy sector. The images in her *Kárahnjúkar Project* (2002–06) show her kissing a large face-shaped rock goodbye before a dam was built, flooding the valley. In her mural *Scheissland* (2005) she presents her critique of the hydropower plant, Kárahnjúkar, as well as the downside to the idealization of pure Icelandic nature through language intervention. The mural, originally exhibited in Germany, consisted of a sarcastic text in German where *Scheisse* (filth/shit) was used as a prefix to many of the words associated with the strategic promotion of Iceland as an untouched wilderness that draws on stereotypical notions of the utopian North.

Artist and politician Hlynur Hallsson also produces artworks that challenge the industrial naturalization of the idea of the wealthy North in narratives of harmless resource extraction. In contrast to Alcoa's promoted brand as an environmentally friendly corporation, his artwork *Drulla-Scheisse-Mud* (2007), a mural in the town of Akureyri, reads, "*Takk fyrir allt álid/vielen dank für das ganze aluminium/thanks for all the aluminum.*" This threefold title in Icelandic (national language), German (a central European language), and the global language of English is another way of allowing language to reflect levels of economic interests and policies that are entwined in resource extraction. The piece became a part of the public space at the height of the Icelandic spending spree. And with its ironic tone, it explicitly mocks Iceland's role as host country for the multinational industry. Thus, Vilhjálmsdóttir's and Hallsson's works pose a pungent critique of national visions for current and future environmental policies and point to their concrete consequences. They challenge the stereotypical ideas embedded in the narratives presenting hydropower as environmentally friendly and clean and thus they send the unpopular message that the political visions of the utopian and wealthy North has consequences beyond serving as an innocent fantasy.

CONCLUSION

In Iceland's current governmental policies, the Arctic is prioritized as the main action area, a field where Iceland can assert its position as a geopolitical agent. The Arctic is a space that hosts the power play of rapidly changing geopolitical structures as well as projections of ecological imaginaries. In art, branding, and politics, one can see responses to the political turmoil in Iceland through the creation of frameworks for imagining the future. The spheres and agendas differ, but at the same time they are connected in their effort to communicate and envision scenarios. In official policies resource extraction and melting polar ice in the Arctic region are factors that are said to enable the future; here, environmental policies seem subordinate to economic policies. The Arctic is currently being portrayed as a new global center and, at the same time, as a frontier region. In negotiations about actions in the Arctic, it is important to remain aware that nation brands and statements, narratives, and images may become a smokescreen that has the potential to redirect attention from the actual ventures. Despite statements about international cooperation and common challenges, the Arctic is not least a space of economic interest and nation-building. The artworks discussed here point to pitfalls in what I have called the Icelandic trinity of purity, harmless energy production, and national image (2014b). Official narratives in branding, politics, and the energy sector support the naturalness of the national project while at the same time naturalizing the extraction and utilization of natural resources.

In an economy based on tourism and exports that rewards images associated with the stereotypes of the utopian, original, and wealthy North alongside political initiatives that focus on resource extraction and national economic benefits of climate change, Iceland benefits from its unique image. The centuries-long tradition of associating Iceland with the original North and the pristine wilderness may serve as a diversion that conceals exploitive practices. By focusing on national and planetary levels, the artworks discussed here indirectly point to a potential ecological risk in relation to policies concerning the future of the Arctic region. However, the contributions of these artists and other critics still only make out a tiny island in the ocean of the consumer, fisheries, and tourism-based economy. The official Icelandic discourses of branding and Arctic policies may thus be understood through an unorthodox application of Mary Louise Pratt's term "anti-conquest," a strategy used to secure an air of innocent intent while asserting immunity against criticism and thus securing the desired hegemony or goal (see also Pratt 1992). This potential was even addressed explicitly by the president in 2005: "no one is afraid to work with us; people even see us as fascinating eccentrics who can do no harm and therefore all doors are thrown wide open when we arrive" (Grímsson 2005, 5). That strategy is enabled by the use of stereotypes of the North that have settled in the cultural conscience both in and outside Iceland. The criticism found in the artworks mentioned here thus becomes an attempt at deconstructing this anti-conquest narrative.

In Iceland, the discourse of Arctic optimism as a way of imagining the future is linked with the conflict around political efforts to deal with the collapse and the resulting crisis by means of dissolving issues of responsibility of individual officials and the anger and anxiety of individual Icelanders into an abstract wound at a national level that is to be healed through a form of solidarity that borders resignation. The Arctic is a dreamscape for negotiations of planetary and national issues, but the main framework for concepts of nature in the official discourse remains focused on utilization. Unlike environmentalist Arne Næss, who proposed a theory about local actors being sensitive to the value of life forms, and the nation state as being the best platform for environmental responsibility (1973, 98; see also chapter "The Polar Hero's Progress: Fridtjof Nansen, Spirituality, and

Environmental History" in this volume), I see the economic interest in the utility of nature that is closely linked with the competition of the nation states in the international market as posing a serious threat to this sensitivity. This problem is reflected in a number of official statements presented in this chapter and not least in the declaration of policy from the current government (Declaration of Policy, Umhverfismál), where the chapter on the national environmental policy concludes that on the national level environmental protection and utilization are two sides of the same coin. The issue of sustainability in the utilization of natural resources, however, is mentioned with reference to the international community. Once more the management and utilization of resources is treated as a national concern, while environmental responsibility is primarily seen as an international issue—the schism that Hallsson's work *Drulla-Scheisse-Mud* refers to.

The artworks mentioned here cut through the naturalizing discourses that support the anti-conquest strategies linking ideas of Icelandic purity with the Arctic policies, and they also expose attitudes based on stereotypes such as the original, wealthy, and utopian North. Some of the works seem to encourage us to ask whether the crisis discourse has supported a state of emergency, a strong focus on growth, and the improvement of international recognition that has allowed Icelandic decision makers to avoid thorough discussions of the long-term consequences. When it comes to environmental issues, these artists direct their criticism at the national level, where questions of responsibility are hard to evade. The artworks call for an in-depth discussion of value systems and of deep versus shallow ecology that is crucial for the dialogue about the future of the Arctic region. Through eco-critical art the ethical dimensions of visions for "Arctic Icelandic" become clearer. Focusing on the environment of the High North in a combined planetary, international, and local framework may very well lead to further fruitful discussions of the future of the Arctic as physical environment rather than as political dreamscape.

This article is based on research conducted as part of the project "Denmark and the New North Atlantic," funded by the Carlsberg Foundation.

WORK CITED

Alcoa Inc. 2013. Sustainable development. Alcoa website. Accessed 16 Nov. http://www.alcoa.com/iceland/en/info_page/sustainable_development.asp

Axelsson, Ragnar, and Mark Nuttall. 2010. *Last days of the Arctic*. Reykjavík: Crymogea.

Branding Report/Iceland's Image. 2013. http://www.ferdamalastofa.is/static/
files/upload/files/200848103017imynd_islands.pdf. Accessed 11 Oct 2013.
Danish Architecture Center. 2013. Greenland – head for the centre of the world.
Danish Architecture Center website. Accessed 13 Nov. http://www.dac.dk/
da/dac-life/udstillinger/2013/groenland---saet-kurs-mod-verdens-centrum/
Erlingsdóttir, Íris. 2009. Changing Iceland's culture. *Iceland Review* 47(4):
66–75.
Gremaud, Ann-Sofie. 2014a. Iceland as center and periphery. Post-colonial and
crypto colonial perspectives. In *The post-colonial North Atlantic: Iceland,
Greenland, and the Faroe Islands*, eds. Lill-Ann Körber and Ebbe Volquardsen,
83–104. Vol. 20 of Beiträge zur Skandinavistik. Berlin: Nordeuropa-Institut
der Humboldt-Universität.
Gremaud, Ann-Sofie. 2014b. Power and purity: Nature as resource in a troubled
society. *Environmental Humanities* 5: 77–100. http://environmentalhuman-
ities.org
Grímsson, Ólafur Ragnar. 2005. How to succeed in modern business: Lessons
from the Icelandic voyage (speech), May 3. Accessed 1 Apr 2014. http://www.
forseti.is/media/files/05.05.03.walbrook.club.pdf
———. 2014. New Year address (speech), January 1. Accessed 18 Jan 2014.
http://www.forseti.is/media/PDF/Aramotaavarp_2014_enska.pdf
———. 2015. The West Nordic dimension in the global Arctic (speech), August
11. Accessed 26 Aug 2015. http://www.forseti.is/media/PDF/2015_08_11_
Faereyjar_Vestnorraena_enska.pdf
Hafstein, Stefán Jón. 2011. Rányrkjubú. *Tímarit Máls og Menningar* 72(3):
6–23.
Hálfdanarson, Guðmundur. 1999. 'Hver á sér fegra föðurland' stada náttúrunnar
í íslenskri þjóðernisvitund. *Skírnir* 173: 304–336.
———. 2000. Þingvellir: An Icelandic 'Lieu de Mémoire.' *History and Memory*
12(1): 4–29.
———. 2011. University of Iceland. A citizen of the *Respublica Scientiarum* or a
nursery for the nation. In *National, Nordic or European?: Nineteenth-century university jubilees and Nordic cooperation*, ed. Pieter Dhont, 285–312. Leiden: Brill.
Huijbens, Edward H. 2011. Nation-branding: A critical evaluation. Assessing the
image building of Iceland. In *Iceland and images of the North*, eds. Sumarliði
Ísleifsson and Daniel Chartier, 553–582. Presses de l'Université du Québec.
Iceland Naturally. Accessed 22 May, 2013. http://www.icelandnaturally.com/
nature/
Icelandic Government. 2013. Declaration of policy. Umhverfismál. http://www.
stjornarrad.is/Stefnuyfirlysing/#umhverfi. Accessed 23 Sept 2016.
Icelandic Government. 2007. Breaking the ice: Arctic development and maritime
transportation, March 27–28 (schedule of conference proceedings). Akureyri:

Iceland. Accessed 10 Sept 2013. http://www.utanrikisraduneyti.is/media/
MFA_pdf/Dagskra_-_Breaking_the_Ice.pdf
———. 2009. Iceland in the Arctic. Arctic report /Ísland á norðurslóðum (website). Accessed 2 Mar 2014. http://www.utanrikisraduneyti.is/media/
Skyrslur/Skyrslan_Island_a_nordurslodumm.pdf
———. 2014a. Declaration of Policy/Stefnuyfirlýsing ríkisstjórnar
Framsóknarflokksins og Sjálfstæðisflokksins (website). Accessed 10 Mar.
http://www.stjornarrad.is/Stefnuyfirlysing/
———. 2014b. Norðurslóðir. Ministry for Foreign Affairs (website). Accessed 2
Mar.   http://www.utanrikisraduneyti.is/verkefni/althjoda-og-oryggismal/
audlinda-og-umhverfismal/nordurslodir/
Ingimundarson, Valur. 2011. Territorial discourses and identity politics. Iceland's
role in the Arctic. In *Arctic security in an age of climate change*, ed. James
Kraska, 174–190. Cambridge: Cambridge University Press.
Ísleifsson, Sumarliði. 2011. Introduction. In *Iceland and images of the North*, eds.
Sumarliði Ísleifsson and Daniel Chartier, 3–22. Presses de l'Université du
Québec.
Law on environmental protection *Altingi*. Accessed 21 Sept 2013. http://www.
althingi.is/altext/stjt/2013.060.html, XII. 69–71 Web.
Lehmann, Orla. 1832. [Review] Om de danske Provindstalstænder med specielt
hensyn paa Island" af B. Einarsson. Cand. jur. Kjøbh. Hos Reitzel. 8. VI. 40
Sider. *Maanedskrift for Litteratur* 523–537.
Næss, Arne. 1973. The shallow and the deep, long-range ecology movement. A
summary. *Inquiry: An Interdisciplinary Journal of Philosophy* 16(1–4): 95–100.
Orkuveita Electricity (digital film). 2013. Orkuveita Reykjavíkur. Accessed 1 July.
http://fraedsla.or.is/Raforka/
Pálsson, Gísli, Bronislaw Szerszynski, Sverker Sörlin, John Marks, Bernard Avril,
Carole Crumley, Heide Hackmann, Poul Holm, John Ingram, Alan Kirman,
Mercedes Pardo Buendía and Rifka Weehuizen. 2013. Reconceptualizing the
'anthropos' in the Anthropocene: Integrating the social sciences and humani-
ties in global environmental change research. In *Responding to the challenges of
our unstable earth (RESCUE)*, ed. Jill Jäger, special issue, *Environmental
Science & Policy* 28: 3–13.
Pratt, Marie Louise. 1992. *Imperial eyes: Travel writing and transculturation*.
New York: Routledge.
Progressive Party Website. http://www.framsokn.is/news/vegna-upphlaups-um-
loftslagsmal-og-matvaelaframleidslu/. Accessed 22 Feb 2016.
Schram, Kristinn. 2009. The wild wild North: The narrative cultures of image
construction in media and everyday life. In *Images of the North: Histories, iden-
tities, ideas*, ed. Sverrir Jakobsson, 249–260. Vol. 14 of Studia Imagologica.
Amsterdam: Rodopi.

Stephens, Piers H.G.   2000. Nature, purity, ontology. *Environmental Values* 9: 267–294.

The Faroe Islands – a nation in the Arctic. Opportunities and challenges. 2013. *Prime minister's office, the foreign service.* Accessed 8 Nov 2013. http://www. mfa.fo/Files/Filer/fragreidingar/101871%20Foroyar%20eitt%20land%20 %C3%AD%20Arktis%20UK.pdf

The Nordic countries. The next supermodel. 2013. *The Economist,* February 2.

Thorhallsson, Baldur, and Christian Rebhan. 2011. Iceland's economic crash and integration Takeoff: An end to EU scepticism? *Scandinavian Political Studies* 34(1): 53–73.

Trans-Arctic Agenda: Challenges of Development, Security, Cooperation. 2013. Arctic portal (website), March 15. Accessed 22 May 2013. http://www.arcticportal.org/ news/25-other-news/969-trans-arctic-agenda-challenges-of-development- security-cooperation

CHAPTER 13

Feminist and Environmentalist Public Governance in the Arctic

Eva-Maria Svensson

The Arctic Council is one of the most important intergovernmental organizations established as a vehicle of public governance in response to the growing interest in the economic potential of natural resources in the Arctic region. The impact of the extraction of natural resources can have severe consequences for climate change and for the living conditions of animals and people in the region. Public governance of the Arctic has many and often contradictory interests to secure, and must seek to balance the economic interests of nation states and private corporations with the interests of the people living in the region and of the environment.

With regards to the question of how the interests of people living there are taken into consideration, it is reasonable to ask whether women and indigenous groups are equally represented in this governance when it comes to setting the agenda for international cooperation. Because these intergovernmental bodies act on behalf of nation states, it is reasonable to expect them to be committed to taking appropriate measures in order to achieve gender equality and promote sustainable development, just as nation states all over the world are obliged to do according to political and

E.-M. Svensson (✉)
Department of Law, School of Business, Economics and Law, University of Gothenburg, Box 650, SE-405 30 Göteborg, Sweden

© The Author(s) 2017
L.-A. Körber et al. (eds.), *Arctic Environmental Modernities,*
DOI 10.1007/978-3-319-39116-8_13

legal commitments. The legitimacy for public governance and the exercise of power relies on democratic values such as gender equality, accountability, transparency, and the representation of *all* citizens.

In this chapter I will focus on gender equality and to some extent sustainability, and the question of whether or not these issues are taken seriously within the public governance of the Arctic.[1] I will explore and analyze how the political and legal obligations designed to achieve gender equality and sustainability are represented within the public governance of the Arctic, and particularly within the Arctic Council. (See Nord 2016a and 2016b for a general analysis of the governance within The Arctic Council.) Public governance produces and reproduces representations of the Arctic and of the people who live there. Through studying the performativity of governance, it is possible to make visible the underlying presumptions of gender equality in order to evaluate them and to (re)imagine a more gender-equal and ecologically sustainable form of governance.

THE THEORETICAL AND CONCEPTUAL FRAMEWORK

The term "public governance" refers to the administrative and process-oriented elements of governing within public bodies, and is synonymous with concepts like "public administration" as the implementation of government policy, or "the organization of government policies and programs as well as the behavior of officials (usually non-elected) formally responsible for their conduct" (UN Economic and Social Council 2006). The concept of public governance encompasses the exercise of power in both an implicit and an explicit way. The implicit aspect is especially of interest when studying the representations of public governance of the Arctic in relation to gender equality and ecological concerns. What is not said is often as important as what is said.

The public governance of the Arctic can be seen in a context in which governance is performed to a great extent by inter- and trans-governmental bodies, by bodies consisting of both governmental and non-governmental organizations, and by bodies consisting of private or semi-private corporations. There is a risk that governance, to an increasing degree, is carried out by administrative bodies far from the citizens of the Arctic. This situation raises questions of transparency, participation, review, and accountability (Cassese 2005; Kingsbury et al. 2005; Reichel 2014). The principle of public access to official records is important for transparency and is a prerequisite for the review and accountability that are necessary components of participatory democracy and broad representation. However, such access becomes more complicated when public governance is transferred to semi-public bodies and to bodies with many nation states involved, including nation states with different or even diverging obligations and ambitions when it comes to gender equality and sustainable development.

Within representations of the Arctic produced by public governance, ecological concerns are addressed as consequences of economic activities such as shipping, oil and gas extraction, and in terms of the impact these activities have on nature and wildlife. The economic activities are considered to be "natural" activities and are not questioned as such. Studying how public governance of the Arctic is performed, imagined, and communicated can contribute to a better understanding of why gender equality and the interests of people living in the area seem to be considered secondary, reactive concerns instead of primary, proactive concerns, as one might expect from the political and legal obligations described above.

My analysis will employ a theoretical framework grounded in a critical branch of legal scholarship called gender (or feminist) legal studies (Gunnarsson et al. 2007; Gunnarsson and Svensson 2009), examining the interrelations between law and policy, and the relationship between "law in books" and "law in action."

In this critical legal tradition, the law is not neutral but is the result of acts of power, and therefore should not be studied as a coherent system, because law is not always rational (Lacey 1998, 5–12). In the Scandinavian context, this critical framework focuses on the power structures *between* men and women, and particularly on redistribution, as opposed to recognition, identity, or differences (see Fraser 1995). This focus on redistribution correlates with the special role that law has as a tool for social change in the Nordic countries. Consequently, there is a close connection between law and policy, and this connection is a focal point in gender legal studies, forming a starting point for the critical analyses of law.

The legitimacy of public governance is based on how well it fulfills its democratically agreed objectives, such as those of gender equality and sustainability. The field of gender legal scholarship studies how law and policy are related with respect to these legal and political commitments and asks whether the objectives are reached, and if not, why. When it comes to gender equality, the Nordic countries project an image of having far-reaching ambitions and of "being nearly there," as having almost reached the goal of gender equality, consistently ranking at the top of the gender equality of the World Economic Forum. The Scandinavian countries also have a quite positive self-image regarding ecology and sustainability.[2] It is therefore worth studying how these Nordic self-images are negotiated in a cooperative intergovernmental body such as the Arctic Council, in which countries with a less positive image in terms of gender equality and sustainability are active.

Public Governance of the Arctic

The Arctic is defined from a variety of aspects and interests and its demarcation is an act of power. The administrative demarcation of the area determines who has the power to govern and over what. Demarcations of the region impacts which states and groups of people have the right to claim the resources and how to respond to claims in the region by private corporations. Several states and private corporations claim access which often contradict the claims made by people living there.

The composition of the Arctic Council and the distribution of decision-making power within it is important when it comes to how different interests are negotiated. Only the member states have decision-making power, and all decisions must be made by consensus. The Council is made up of eight member states (Canada, Denmark including Greenland and the Faroe Islands, Finland, Iceland, Norway, the Russian Federation, Sweden, and the USA) and six permanent participants, who represent indigenous people and include the Arctic Athabaskan Council, the Aleut International Association, the Gwich'in Council International, the Inuit Circumpolar Council, the Russian Association of Indigenous People of the North, and the Sámi Council. Permanent participants have full consultation rights in connection with the Council's negotiations and decisions, a status available to Arctic organizations of indigenous peoples representing a single indigenous group spanning more than one Arctic state or multiple indigenous populations in a single Arctic state.

The Council also has observers, a status open to non-Arctic states, to global and regional intergovernmental and interparliamentary organizations, and to non-governmental organizations. There seems to be great interest in becoming an observer in the Council, and, as of July 2014, twelve non-Arctic states, nine intergovernmental and interparliamentary organizations, and eleven non-governmental organizations had been given observer status.

According to the Council, its goal is to provide a valuable platform for discussion on all issues relevant to the Arctic and the people who live there. These include environmental protection, climate change, Arctic and circumpolar biodiversity, marine and shipping activity in the oceans, and the welfare of Arctic peoples. Specifically regarding peoples, the focus is on health and wellbeing, as well as on the preservation of cultural heritage and language. The eight member states in the Council, along with the EU, have adopted their own national strategies for the region in recent years with somewhat differing priorities. What they all have in common is the highlighting of environmental issues, but not gender equality. For example, the primary focus of the Swedish strategy in the region is climate change (Arctic Secretariat 2011).

Gender Equality and Sustainable Development as Goals and Obligations for Public Governance

The objectives of gender equality and sustainable development are related to each other in practical, political, and theoretical ways. Statistical data shows that men tend to live in a way that has a greater impact on the climate and have larger ecological footprints than women. Men also dominate the political system, especially in many ecologically relevant areas such as transport (Svedberg 2014, 24). In feminist ecotheory, domination over nature has been connected to domination over women (Warren 2008; Merchant 1980). If there is indeed such a connection, can it be found in the representations of public governance of the Arctic? The rhetoric within public governance is quite ambitious when it comes to ecological concerns, but the contrary can be said when it comes to gender equality. Does this mean that the connection between the two fields of domination is not prevalent in the Arctic, or does it mean that the ecological rhetoric is simply rhetorical?

Gender equality, or equality between women and men, is regarded as a human right according to The Universal Declaration of Human Rights (1948). Gender equality is a *legal obligation* for most states in the world as well as a political goal for many others. Sweden is an exemplary case, and has had a specific policy area called "gender equality" since 1972 with far-reaching objectives that go beyond international legal obligations. The strategy of "gender-mainstreaming" as a method for public governance was adopted in 1994, a year before it was adopted by the UN as a strategy for obtaining gender equality throughout the world in the 1995 Beijing Declaration and Platform for Action (United Nations Entity for Gender Equality and the Empowerment of Women 1995). According to the

World Bank, a "central element of good governance is the responsiveness of policies and public institutions to the needs of all citizens. Policies and institutions must represent the interests of women and men and promote equal access to resources, rights, and voice" (World Bank 2006; see also United Nations 2000).

One of the best-known documents stipulating the ramifications of sustainable development is the Brundtland Report (United Nations 1987), which defined sustainable development as a "development that meets the needs of the present without compromising the ability of future generations to meet their own needs." Five years later, in 1992, the UN Conference on Environment and Development published the Earth Charter and adopted an action plan, Agenda 21. Integration of environmental and social concerns into all development processes was stressed as essential, in combination with an emphasis on broad public participation in decision making. Since the publication of the UN document, the meaning of the concept "sustainability" has come to include also cultural sustainability, in addition to economic, environmental, and social sustainability. The Millennium Declaration (United Nations 2000) identifies the three principles of economic development, social development, and environmental protection. In the post-2015 agenda of Sustainable Development Goals (UNDP 2015), sustainability is an overall objective for all 17 goals. Sustainable development is also an overall objective within the EU, as expressed in the Treaty of Amsterdam in 1999.

The full development and advancement of women for the purpose of guaranteeing them the exercise and enjoyment of human rights and fundamental freedoms on the basis of equality with men in all fields—especially in the political, social, economic, and cultural fields—goes hand in hand with broad public participation in decision making as a fundamental prerequisite for achieving sustainable development. An important question to ask is whether the public governance of the Arctic fulfills these objectives.

Public governance relies on and is expected to fulfill the ambitions expressed in political and legal documents. The obligations expressed in such documents can, as mentioned above, require more or less action on the part of the nation state. The more legally binding the obligation, the more it can be expected that the desired outcome and results will be achieved; and, if they are not, then stronger criticism can be directed toward the governing bodies. However, the governance as such can be more or less active. There is always, of course, a certain level of discretion for how public governance bodies are organized as well as how they carry out their duties. Thus public governance can be characterized as *reactive* or *proactive* when it comes to the fulfillment of its obligations and when it comes to prioritizations. To (only) follow the legally binding obligations can be considered reactive, and to use the discretion to take affirmative or positive action and go further can be considered proactive.

REPRESENTATIONS OF PUBLIC GOVERNANCE IN THE ARCTIC

How is the public governance of the Arctic represented and what does this representation say about public governance? The governance of the Arctic is structured around four topics: environment and climate, bio-diversity, oceans, and Arctic peoples. The first images a visitor sees on the homepage of the website of the Arctic Council are of animals and representations of indigenous peoples. A polar bear represents the topics of environment and climate, a whale represents the oceans, an arctic fox represents biodiversity, and two children playing in the snow represent the Arctic people. Exploring the different topics on the website and reading about the many working groups with their special responsibilities gives one the impression that this is a governing body that is concerned about the environment and people. These images of smiling people, beautiful animals, and the environment merge together with the imaginary of meetings and discussions to give the impression of problem-solving governance without many conflicts. The organization of the governance seems to be based on mutual understanding and respect between the people of the Arctic (mainly visualized as indigenous) and the states with interests in the region. However, as already mentioned, indigenous groups have only "full consultation rights" in connection with the Council's negotiations and decisions, not the right to make such decisions. The right to decide lies in the hands of the eight member states together. The question of who represents the member states is, of course, a relevant one. Are there indigenous people among the representatives for the member states? And what is the proportion of women and men?

It is obvious that the main focus of the governance is on the first three of the four topical areas. For example, during the chairships of Norway, Denmark, and Sweden between 2006 and 2013, the common objectives were climate change, environmental protection, circumpolar observation, monitoring of change in the Arctic, integrated management of resources, indigenous people, and local living conditions—listed in order of their priority. The economic development of the Arctic seems to be a self-evident focus of the Council's work and is not questioned as such. The main objective for Canada's chairship in 2013–15 was to promote economic development, despite the fact that the latest Declaration of the Arctic Council from 2013 explicitly stresses the need to improve both the economic and social conditions of indigenous populations. Business is given a special role in the development of the Arctic, and the Council intends to increase the cooperation and interaction with the business community as a way to ensure sustainable development in the region (Kiruna Declaration 2013). One example of this is the Circumpolar Business Forum, created under the Canadian chairship, of which one of its objectives is "to provide a business perspective to the work of the Arctic Council" (Arctic Economic Council 2014). The US chairship (2015–17) has formulated three priority areas: to improve a) economic and living conditions in Arctic communities; b) Arctic ocean safety, security, and stewardship; and c) the impact of climate change. (U.S. Department of State 2015).

GENDER EQUALITY AND FEMINISM ACCORDING TO THE ARCTIC COUNCIL

Until now, the nine declarations of the ministerial meetings, held every second year from 1998 to 2015, have been signed by 57 (76 %) men and 18 (24 %) women (www.arctic-council.org). Of the many working groups, in 2013 only one—the Sustainable Development Working Group (SDWG)—was chaired by a woman (in 2016 the numbers have become even). Human health and socio-economic issues are the focus of SDWG's major projects. The most important document produced by the group are the two Arctic Human Development Reports (AHDR 2004, 2015). The first AHDR report was published during the Icelandic chairship in collaboration with other bodies such as the United Nations Development Program in order to initiate the development of a knowledge base for the Arctic Council's Sustainable Development Program. The scope of the report was broad and it covered demographics, core systems (including societal, cultural, economic, political, and legal systems), and so-called cross-cutting themes.

"Gender issues" are considered in a special chapter (11) in the cross-cutting themes section of AHDR 2004. The chapter addresses several critical issues but does not provide an overall assessment of gender issues in the Arctic. It does attempt to explain different notions of feminism and how feminism is viewed within specific communities, contrasting "Western feminism" with a "non-Western or an indigenous feminism," with these concepts presented as singular and coherent. The differentiation of feminism into these two categories is problematic for several reasons.

The relation between "feminism" and "gender equality" is not explicitly discussed in the report at all. Feminism is, generally speaking, a political movement for women's rights with different political tendencies (such as liberal, conservative, radical, socialist). Feminism is also a research perspective with a common emancipatory interest in knowledge, encompassing different models which explain why gender inequality is prevalent. The explanations can be individual or structural, and can focus on redistribution of both power and resources, or on identity and recognition (see Fraser 1995).

Gender equality, on the other hand, is an objective anchored in legal documents on several levels and is presented as an obligation for nation states to strive for (at least the states that have ratified certain legally binding documents such as The Convention on the Elimination of All Forms of Discrimination against Women from 1979 [CEDAW]). Gender equality has also been broken down into indicators that can be measured and compared in different kinds of gender equality indexes. These indexes tend to identify different material conditions such as education, political and economic power, and health.

The discussion on different kinds of feminism in AHDR 2004 is far removed from the legal concept of gender equality. The report assumes that gender equality and Western feminism (i.e., liberal feminism) are synonymous (Stefansson Arctic Institute 2004). It is also notable that CEDAW, which is the legally binding document ratified by most states in the world, is not mentioned in the report as a benchmark for defining gender equality. CEDAW prohibits discrimination of women and calls for action by nation states to eliminate discriminatory practices and actions, which is far from what is considered Western feminism in the report.

The lack of focus on gender equality in favor of a focus on "feminisms" opens the way for positioning Western feminism as oppositional to the interests of indigenous women and as critical of traditional living conditions. The CEDAW talks about legal non-discrimination and independence, not of banning certain tasks for each sex. Even though the need for "defining power relationships" is mentioned in AHDR 2004 (although without signifying any particular power relationships), it seems to be understood that gender equality is opposed to traditional gender roles, or even that it suggests new ways of organizing indigenous people's lives, which is expressed with the example of the man being at home and the woman participating in the labor market. Gender equality according to the political and legal perceptions is more about equal values, rights, duties, and power in both private and public life, and not about everybody doing the same thing. Even though independence is highly valued and maintained in (Western) liberalism—and as such is questioned in the report as contradictory to a traditional way of living—independence protects people from exploitation, abuse, and discrimination, which is just as important in a traditional setting as in a "Western" setting. Independence does not mean that people are not mutually independent from each other. Gender equality as a legal principle can be understood as protecting individuals from negative dependence (such as when somebody is not able to leave a relationship if he or she wants to) and encouraging positive dependence (choosing to live in relationships based on free will and not coercion).

Implicitly seeing Western feminism as synonymous with gender equality and giving it a certain meaning in opposition to traditional or new ways of living in indigenous groups could actually strengthen the dichotomy between indigenous and non-indigenous groups. The following quotation is an example of such an implicit presumption: "today, one might find among younger couples a situation where a mother holds a job outside the home while the husband is the homemaker with three or four children at home. … These observations demonstrate that gender equality issues have to be understood from a uniquely Arctic perspective, different from the typical idea of power imbalance between males and females" (AHDR 2004, 189).

Men's changing roles in Arctic society and how they affect social problems are highlighted in the report. There is a devaluation of men's traditional roles, and their welfare is seen as being much more jeopardized and at risk than that of women. In fact, the report describes how modern development in the Arctic is "systematically disenfranchising Arctic men" (AHDR 2004, 191). This is said to be in contrast to the assumptions of Western general feminist discourses in which women's situations are supposedly worse than that of men. This assertion is not in accordance with the gender equality discourse in the EU and Sweden, where the position of men in modern society is still a highly topical issue, as when boys perform worse than girls at school—an issue which is addressed as a severe problem. Second, the presumption seems to be that if men's outcomes are getting worse it must be that women's outcomes are getting better. This is not always the case, however, and there might be other groups of men that are gaining power and influence at the expense of those men who lose their power and influence. Third, men's social problems do not only have impacts on men, but often also on women; and male violence against women seems to become worse when men are devalued.

The report also provides some empirical data on women and men in the Arctic. There is a pattern of disproportionate migration out of the region by young women, resulting in a population made up of predominantly young adult and middle-aged males in many places. Education seems to be a major reason why women leave, but there are also complex relationships between individual and structural "push and pull" factors that influence this migration. Many of the factors mentioned in the report seem to be related to a lack of influence and power. Additionally, the kinds of activities, jobs, educational opportunities, and future scenarios in the region seem to attract more men than women. Women's higher rate of marriage to "outsiders" plays a significant role in their migration, according to the report (AHDR 2004, 192). Women seem to act as if they were better off leaving the region as compared to men. The problem is highlighted in the report as a matter of concern for public governance. Until now, the problem seems not to have been taken seriously as a part of any gender equality policy for the public governance of the Arctic. The situation seems to be seen as a natural consequence of the development in the region and, therefore, is not directly questioned.

There are several more issues raised in the report. It might be expected that the Arctic Council would have taken the issues raised about gender equality more seriously, but this has not been the case. The second Arctic Human Development Report (AHDR 2015) was released in 2014 (see Larsen and Fondahl 2015). In this report, gender equality is mainstreamed. Gender mainstreaming is a globally accepted strategy for promoting gender equality, adopted by the UN in 1995 as part of the Beijing Declaration and Platform for Action (United Nations Entity for Gender Equality and the Empowerment of Women 1995). The strategy has its pros and cons, and there is a risk that it might make gender equality invisible and unreflective. The 2015 report states that there is a lack of knowledge of gendered dimensions in several aspects, and that gender equality is a pre-requisite for human development, well-being, and dignity in the Arctic. Any difference between the 2015 and 2004 AHDR in this respect is thus more or less negligible.

Toward a Gender-Equal and Ecologically Sustainable Public Governance

AHDR 2015 is an example of what can be understood as the strengthening of a gender-equal and ecologically sustainable public governance of the Arctic. The explicit underlying assumption behind the report is that sustainable development is a human-centered concept (Larsen and Fondahl 2015). Given the nature of northern societies, economies, and environments, there should be particular emphasis on human–environment relations in the Arctic as well as on the ties between individual well-being and the health of northern communities. The concept of sustainable human development reinforces the human dimension of sustainability and has been suggested as a way to put priorities into perspective and to stress the importance of human wellbeing as the ultimate goal of sustainable development. The connection between gender equality and ecological concerns seems to be essential in this framing. Sustainable human development must, as one of the expectations of public governance, be based on and take into account the interests of *all* peoples living in the region.

The explicit emphasis on human–environmental relations within the public governance of the Arctic goes together with theories developed in gender legal studies and in feminist ecotheory. According to these theories, there seem to be connections between changes in society and how nature and the relations between men and women are perceived. The relationship between human beings and nature seems to have similarities with the relationship between men and women. Translated to the realm of Arctic governance, this means either that a lack of focus on gender equality goes hand in hand with a perception of nature as something that can be controlled and exploited by human beings (usually men) or that the focus on gender equality goes hand in hand with a perception of nature as an organic process that interacts with human beings.

The quest for sustainable development and the quest for gender equality seem to share many of the same characteristics, and these are brought together in the concept of a gender-equal and ecological view of the world. Leaving the development of important democratic values such as gender equality and sustainability to different actors without supporting the process with efficient and compulsory means might result in a situation in which the values are not taken seriously. This risk is even more obvious when the values are in conflict with other interests such as economic development and the extraction of natural resources. One might expect that such conflicts of interest would lead to debates on how the interests should be weighed in relation to each other, but this does not seem to be the case.

Public governance as such, and the text of the AHDR 2004 in particular, both give a picture of development that goes smoothly and without any significant conflicts. The lack of a conflict is significant in that the impact of the extraction of natural resources on the environment and people are not addressed in terms of gender equality by the Arctic Council.

Much of the extraction business is performed by multinational corporations, and thus the extraction of resources does not necessarily benefit the people within the region. The extraction industry attracts mostly men, and not always from the region. Fly-in and fly-out patterns are increasing, which have led to new problems in the region, particularly human trafficking and prostitution. Mobility patterns within the region also have impacts on families and on the relationships between men and women. Most of these problems are highlighted in AHDR 2004. However, the perceptual framework for these problems seems to be reactive and not proactive. The human and environmental dimensions are problematized, but not the economic dimension, with economic interests in the region taken as a given.

The human dimension, even though it is one of the top priorities in the public governance of the Arctic, is perceived as *reactive* to other top priorities such as economic development and climate and the environment. Economic development and climate change, which are effects of political and economic prioritizations and the way we live, seem to be considered as processes that occur naturally, without human intervention. The human dimension is subordinated; which says two things. First, no one is identified as responsible for the development of the people in the region, and especially not for the so-called vulnerable groups such as indigenous people and women. Second, activities and prioritizations by states representing majority populations affect indigenous populations and other groups like women, but these activities and prioritizations are not questioned as such within the governance of the Arctic. The power structure of a culture of economic development dominating over nature seems to go well with non-indigenous people dominating over indigenous people and men dominating over women. The historical connection between the domination of nature and the domination of women seems to be confirmed within the realm of public governance of the Arctic. This comes despite lofty ambitions and explicitly stated objectives to promote sustainable development—which according to AHDR 2015 is a sustainable *human* development—and environmental protection in the Arctic. There seem still to be a few steps left to take before we see *active* governance that is actually based on a gender-equal and ecologically sustainable worldview.

Notes

1. The chapter is written as part of an international cross-disciplinary comparative research program, TUAQ, with scholars from law, economical statistics, and political science with competence in gender studies. The focus of the program is on public mechanisms for gender equality in the Arctic.
2. Note that there is a discrepancy between the self-image, the image, and what is statistically borne out in the data. When comparisons are made between different countries, the Nordic countries top the statistics both for Europe and the rest of the world when it comes to ecological footprints (Equal climate, www.equalclimate.org/en/consumption/fascinating_figures/. Accessed 24 Sept 2016).

Work Cited

AHDR. *Arctic Human Development Report 2004.* Akureyri: Stefansson Arctic Institute.

Arctic Economic Council (AEC). 2014. Arctic economic council. http://www.arctic-council.org/index.php/en/our-work2/8-news-and-events/195-aec-2. Accessed 20 Sept 2016.

Arctic Secretariat. 2011. *Sweden's strategy for the Arctic region, ministry for foreign affairs, department for Eastern Europe and Central Asia.*

Cassese, Sabino. 2005. Administrative law without the state? The challenge of global regulation. *New York University Journal of International Law and Politics* 37: 663–694.

Fraser, Nancy. 1995. From redistribution to recognition? Dilemmas of justice in a 'post-socialist' age. *New Left Review* 212: 68–93.

Gunnarsson, Åsa, and Eva-Maria Svensson. 2009. *Genusrättsvetenskap.* Lund: Studentlitteratur.

Gunnarsson, Åsa, Eva-Maria Svensson, and Margaret Davies. 2007. *Exploiting the limits of law. Swedish feminism and the challenge to pessimism.* Aldershot: Ashgate.

Kingsbury, Benedict, Nico Krisch, and Richard B. Stewart. 2005. The emergence of global administrative law. *Law and Contemporary Problems* 68(3–4): 15–62.

Kiruna Declaration. 2013. *The eight ministerial meeting of the arctic council.* May 15, 2013. Kiruna, Sweden. http://hdl.handle.net/11374/93. Accessed 20 Sept 2016.

Lacey, Nicola. 1998. *Unspeakable subjects: Feminist essays in legal and social theory.* Oxford: Hart Publishing.

Larsen, Joan Nymand and Gail Fondahl, eds. 2015. *Arctic human development report: regional processes and global linkages,* (AHDR 2015). TemaNord 2014: 567. Copenhagen: Nordic Council of Ministers. http://urn.kb.se/resolve?urn =urn:nbn:se:norden:org:diva-3809. Accessed 20 Sept 2016.

Nord, Douglas C. 2016a. *The changing Arctic: Creating a framework for consensus building and governance within the Arctic council.* New York: Palgrave Macmillan.

———. 2016b. *The Arctic council: Governance within the Far North.* New York: Routledge.

Merchant, Carolyn. 1980. *The death of nature:Women, ecology, and the scientific revolution.* San Francisco: Harper & Row.

Reichel, Jane. 2014. Communicating with the European composite administration. *German Law Journal* 15: 883–906.

Svedberg, Wanna. 2014. *Ett (o)jämställt transportsystem i gränslandet mellan politik och rätt.* Malmö: Bokbox Förlag.

The Convention on the Elimination of All Forms of Discrimination against Women (CEDAW). 1979. http://www.un.org/womenwatch/daw/cedaw/. Accessed 20 Sept 2016.

UN Economic and Social Council. 2006. *Definition of basic concepts and terminologies in governance and public administration.* Committee of Experts on Public Administration.

UNDP. 2015. *Sustainable development goals.* http://www.undp.org/content/undp/en/home/mdgoverview/. Accessed 20 Sept 2016..

United Nations. 1948. *The universal declaration of human rights.* http://www.un.org/en/universal-declaration-human-rights/. Accessed 20 Sept 2016.

———. 1987. *Report of the world commission on environment and development: Our common future* (Brundtland Report). http://www.un-documents.net/our-common-future.pdf. Accessed 20 Sept 2016.

———. 2000. *United nations millennium Declaration.* A/RES/55/2. http://www.un.org/millennium/declaration/ares552e.pdf. Accessed 20 Sept 2016.

United Nations Entity for Gender Equality and the Empowerment of Women. 1995. *Beijing declaration and platform for action.* http://www.un.org/womenwatch/daw/beijing/platform. Accessed 20 Sept 2016.

Warren, Karen. J. 2008. The power and the promise of ecological feminism. In *Environmental ethics: Readings in theory and application*. Louis P. Pojman, Belmont, CA, Thomson Wadsworth.

World Bank. 2006. *Governance and gender equality*. Gender and Development Group. http://www.capwip.org/readingroom/TopotheShelf.Newsfeeds/2006/Governance%20and%20Gender%20Equality%20%282006%29.pdf

www.arctic-council.org. Accessed 20 Sept 2016.

www.arctic-council.org/index.php/en/about-us/working-groups/aec. Accessed 20 Sept 2016.

www.equalclimate.org/en/consumption/fascinating_figures. Accessed 20 Sept 2016.

CHAPTER 14

The Greenlandic Reconciliation Commission: Ethnonationalism, Arctic Resources, and Post-Colonial Identity

Kirsten Thisted

Today, only the most stubborn imagination maintains the Arctic as a pristine, empty, white space, distant, and dangerous. At this point, everybody ought to know that the Arctic was populated long before the age of the so-called Arctic explorations. International news outlets are constantly flashing headlines about the drastic changes due to global warming, the rush to exploit resources, and the opening of new sea routes. As such, there is nothing new about the important role of the Arctic in international politics. What is fundamentally different today is the status of the peoples of the Far North. No longer can these peoples be governed and treated as voiceless "natives" to be compared with the marine mammals, birds, and fish of the area. Various forms of self-rule are now the norm rather than the exception. This new situation has not emerged without turmoil and tough political negotiations. Old asymmetrical power relations continue to make their influence felt, and past events still need to be

K. Thisted (✉)
Associate Professor Department of Cross-Cultural and Regional Studies, University of Copenhagen, Copenhagen, Denmark

© The Author(s) 2017
L.-A. Körber et al. (eds.), *Arctic Environmental Modernities,*
DOI 10.1007/978-3-319-39116-8_14

dealt with in the present. "Reconciliation" is therefore a relevant term also in an Arctic context.

However, the political situation is often complicated, and it varies considerably from area to area. Different locales have had very different histories because they have interacted with different empires and nation states. Today, the Arctic is inhabited by indigenous peoples, immigrants who have been in the area for generations, more recent immigrants, as well as mixed ethnicities across these groups. The Arctic is and has for centuries been a contact zone, according to Mary Louise Pratt's famous definition: "social spaces where disparate cultures meet, clash, and grapple with each other, often in highly asymmetrical relations of domination and subordination—like colonialism, slavery, or their aftermaths as they are lived out across the globe today" (Pratt 1992, 4).

Indigeneity and the discussion about the rights of indigenous peoples have played an important role in the changing status of such peoples in the Arctic. However, "indigeneity" is not an unproblematic term. Being so closely tied to ideas of culture and origins, the discourse of indigeneity seems bound to promote essentialism and—in cases like Greenland, where the indigenous people are the majority and are in power in a state-like set-up—even ethnonationalism. In this situation, reconciliation is not a clear-cut matter between the former colonizers and the former colonized but instead a rather murky affair with unclear boundaries between the parties in the conflict.

This chapter outlines the background for and public debate about the Greenlandic Reconciliation Commission, which was established in 2014. The intent is to examine the different agendas and positions at the time when the Commission was established. Should the Commission be seen as an act of ethnonationalism? Or as an attempt to put colonialism to rest and make room for a post-post-colonial condition (Gad 2009), where the focal point is no longer ethnicity and the old opposition between Danish and Greenlandic? The conclusion is that both intentions are probably in play, and that it is too soon to determine which is going to come out on top.

No matter what, the Reconciliation Commission proves the importance of including emotions in the analysis of political processes. Since the implementation of Greenlandic Self-Government in 2009, crucial questions have been debated and put to a vote in the Greenlandic parliament: should Greenland open the country to large-scale mining and industrial projects, necessitating the import of a huge foreign workforce? Should Greenland begin extracting uranium from its underground, with all the dangers this poses to the environment—not to mention the security issues connected with such extraction? Questions such as these challenge the power relations between Denmark and Greenland and put to a test Greenland's power to make decisions independently of the Danish government.

Likewise, debates bring to the fore how past relations of supremacy and subordination influence decisions that are made today concerning the future. Economic independence is a prerequisite if Greenland is to achieve full political independence. Thus, the discussion about resources and sub-soil is driven by the desire for a future where ancient inferiority is replaced with pride and equality.

An important task for the Reconciliation Commission is to bring insight into and reconciliation with the socio-historical development in Greenland, including the way in which modernity changed the lives of the Greenlanders. The suggestion is not that the responsibility should rest exclusively with the colonizers but that the Greenlanders too need to reconcile their own involvement in and responsibility for the process. Therefore, insight into the Reconciliation Commission and the way in which it was set up adds important input to the discussion about the right to define and interpret the past—and to take responsibility for the past. This may in due time open new opportunities as concerns the future.

A GREENLANDIC NATION

In Greenland, the discourse of indigeneity is currently being transformed from a language of resistance to a language of governance. Today, Greenland and the Faroe Islands are the only overseas territories left from the once far-reaching Danish empire. Greenland was colonized by the Danish–Norwegian state in 1721. In 1953, Danish colonialism officially ended when Greenland became an equal part of Denmark as the northernmost county. In 1979, home rule was implemented, followed in 2009 by self-government, an expansion of home rule. The Self-Government Act confirms that it is up to the people of Greenland to decide whether and when they might want to withdraw from the commonwealth and achieve full independence. Both Greenlandic home rule and the Act on Greenland Self-Government emerged in a political environment with a strong awareness of indigenous rights. The work to formulate the Act on Greenland Self-Government ran parallel to the United Nations (UN) negotiations on the Rights of the Indigenous Peoples (Kleist 2011; Thisted 2013). However, the term "indigenous peoples" is not mentioned anywhere in the act. It was deliberately avoided, as Greenland would otherwise have remained in a minority position. While the UN Declaration attempts to regulate the relationship between indigenous peoples and states, seeking to protect the former from the latter, the Act on Greenland Self-Government gives Greenland the status of an "equal partner" with the Danish Government. It remains an open question whether indigeneity according to the UN Declaration translates from an ethnic minority position to governance. The issue is also being debated internally in Greenland, where support for the rhetoric about Greenlandic indigeneity has never been unanimous, not least due to the connotations of under-development and suppression—and anti-modernity—that cling to the term. Some politicians have suggested that the term be abandoned after the implementation of self-government (Johansen 2008, quoted in Thisted 2013, 235), while others want to pre-serve the term, not least because it still conveys certain special rights, for instance with regard to whaling. The various forums for indigenous peoples make up an important part of the Greenlanders' international contacts and networks. This includes the connections with the "kinsfolk" in Canada, Alaska, and Siberia, and with the Sámi people in northern Scandinavia. However, not all Greenlanders identify as Inuit—or rather, they iden-tify as various "degrees" of Inuit, since all Greenlanders today number Danes/Scandinavians/Europeans or other non-Inuit among their ances-tors. *Kalaallit* (singular: *kalaaleq*) is the Greenlandic (*kalaallisut*) term that covers this modern, mixed population. Still, anyone who calls him or herself *kalaaleq* can usually claim some relation to the Inuit and the Inuit language. In Greenlandic, Greenland is called *Kalaallit Nunaat*, literally the land (*nunaat*) of the Greenlanders (*kalaallit*).[1]

The Act on Greenland Self-Government is an agreement between Denmark and Greenland, and there is no mention of ethnic groups. However, as men-tioned above, it does mention "the people of Greenland" (2009, §21). In Greenlandic, this is expressed as *inuiaat kalaallit*. Thus, the Greenlandic nation is ethnically defined, in accordance with the concept of the nation state inherited from Denmark, which is based on *ethnos*, the shared heritage and culture of the people (Thisted 2011; Langgård 2011). In recent years, this idea has been challenged and negotiated in politics, art, and popular cul-ture, and a new vision of the nation has emerged, based on a concept of the people as a *demos*: the political community of all the people inhabiting a given territory (Thisted 2012a, b, 2014a, b; Otte 2013). The discussion has clear similarities to the ongoing debate in Denmark about immigration.

Not surprisingly, self-government has sparked debate about what kind of society Greenland wants to become—a debate that is also played out in relation to party politics and the struggle for power. Even though it deals with the past, the Reconciliation Commission is part of this discussion.

The discussion about Greenland's future as either part of the Danish realm or an independent nation state often revolves around arguments for or against the need for (further) decolonization. This makes the field of reconciliation extremely complex, and although the topic has not received much attention or been widely debated, it may bring to the surface some of the issues that have remained unspoken in other debates, chief among them the ongoing discussion about the use of Danish, which for a variety of reasons has remained a language of power, although Greenlandic was in fact declared to be the country's main language with the introduction of home rule in 1979.

RECONCILIATION, APOLOGIES, POLITICS

The idea of a reconciliation commission in Greenland is not new but was proposed in the late 1990s by the Somalian-born psychiatrist Fatuma Ali who worked in Greenland after spending many years in Denmark. In 2004, Ali held the first reconciliation seminar for a group of 18 people in Tuscany (Ali and Lindhardt 2006). Even by then, the initiative met with objections against the perceived comparison between the brutal apartheid regime in South Africa and the non-violent Danish administration of Greenland (Lidegaard 1998; Petersen 1998). The initiative therefore fizzled out to some extent—but the ideas were not forgotten, not least because the question of reconciliation had received so much international attention, including in relation to events that did not involve mass murder and genocide. In 2008 the Canadian Government issued a formal apology to the indigenous population for the country's assimilation policy, which aimed to impart the majority language and culture at the cost of the people's own background. At the same time, a Truth and Reconciliation Commission was established to investigate the Canadian Indian residential schools (active 1876–1996) and make the commission's findings publicly known. Previously in 2008, the Australian prime minister had apologized on behalf of the nation for more than 70 years of forcible removals of aboriginal children from their families, the so-called "stolen generations."

The Greenlandic public has followed these events closely, considering whether Greenland may have similar cases involving possible human rights violations. The situation in Greenland stands out, however, because primary and upper secondary school pupils were taught mainly in Greenlandic—at least until the demand for teachers, combined with a

Greenlandic desire for future generations to learn Danish, led to a massive influx of Danish teachers throughout the 1960s and 1970s. A case of about 22 children who were separated from their families in 1951 and sent to Denmark to be immersed in Danish language and culture bears some resemblance to the cases in Canada and Australia, and some have called for an apology from Denmark—not least after the feature film *The Experiment* (*Eksperimentet*; Louise Friedberg 2010) sparked renewed debate about the issue.

Scandinavia is not new to official apologies. In 1997, the Norwegian king apologized on behalf of the nation to the Sámi people for the oppression and Norwegianization they had been subjected to over the years. The following year, the Swedish state followed suit with an apology that even explicitly mentioned the colonization of northern Sweden. A similar apology from Denmark for colonizing Greenland appears unlikely, however, not least because the narrative about the well-intentioned Danish colonialism is so persistent (Olwig 2003; Thisted 2009; Jensen 2012a, b; see also chapter "Cod Society: The Technopolitics of Modern Greenland" in this book). According to this narrative, Denmark ruled with the consent of the Greenlanders and with their growing political participation (Jensen 2012a, 2012b; Thisted 2012a, 2012b).

The only apology offered by Denmark to Greenland related to the forced relocation of the inhabitants of Thule to make room for the American air base there in 1953. This apology was issued in print in 1999 on behalf of the Danish state and signed by the prime minister of Denmark, Poul Nyrup Rasmussen, and by the Greenlandic premier, Jonathan Motzfeldt, "in the spirit of the commonwealth and with respect for the population of Greenland and Thule." The apology applied to the way in which the relocation was decided and implemented, not to the relocation per se. The case eventually went before the Danish Supreme Court, which ruled in 2003 that the relocation was an expropriation and awarded the inhabitants monetary compensation. (In the plaintiff's opinion, the amount was ridiculously low.)

Throughout the 2000s, the upcoming Self-Government Act was the Greenlandic population's primary concern. The main focus was on the implementation of the act and the emerging debate about a possible Greenlandic constitution—a somewhat controversial topic, since Greenland is still part of the Danish realm. As was the case for the Home Rule Act, the Self-Government Act immediately led to a test of the legal boundaries of independence. Throughout this period, the gaze remained firmly fixed on the future. After the 2013 election, however, reconciliation with the past made it onto the political agenda.

ALEQA HAMMOND AND THE GREENLANDIC RECONCILIATION COMMISSION

The March 2013 election handed the reins to the Siumut Party, which had held the premier's office and thus been the dominant political power throughout the home rule years. Siumut is widely characterized as a Social Democratic party, while the shift to self-government had put the more left-ist party Inuit Ataqatigiit in power under the leadership of Kuupik Kleist, an experienced politician who had also held top administrative jobs in Greenland's home rule administration. The 2013 election gave Greenland its first female premier, the less experienced but charismatic Aleqa Hammond, who had been chairman of Siumut since 2009. Hammond was elected to the Greenlandic Parliament with the biggest number of personal votes ever in a Greenlandic election. To Hammond, decolonization means full independence from Denmark, and after the election she made the issue of independence her leading international issue under the motto "In my life time." Hammond had the issue of reconciliation written into the coalition agreement, and the Reconciliation Commission was subsequently included in the Greenlandic national budget with a four-year annual grant of DKK2.4 million, beginning in 2014—a considerable amount compared to Greenland's gross national product. Since the question of reconciliation was now part of the coalition government's program, it also became the object of criticism from the opposition. The main focus of this criticism was that the money would be better spent elsewhere, and that the issue detracts attention from other, more relevant, issues such as unemployment and the poor state of the economy. In addition, many still found that establishing a reconciliation commission gives a misleading impression of historical Danish–Greenlandic relations due to the implied associations with genocide and South African apartheid.

The main stumbling block, however, has been the rhetoric in which the commission was launched, especially the talk of different "population groups" in Greenlandic society. Hammond devoted a large section of her New Year's address to the topic of reconciliation, including the following comments:

> I would suggest that in the intersection between our history and the rela-
> tions between our population groups we can find some of the greatest

taboos in our country today. We must break down these taboos in order to reconcile ourselves to today's situation and to strengthen our own self-awareness. We must also be ready to change things, should discrepancies be discovered. (Hammond 2014)

This sparked many comments in Greenlandic newspapers and in the social media, as people wondered whether the government wanted to reintroduce ethnic distinctions in society and, if so, how the term "Greenlander" (*kalaaleq*) should be defined. As synonymous with Inuk/Inuit? And what degree of purity would be required to qualify as Greenlandic? Greenlanders who had Danish as their first language expressed concern about the impact of this form of "reconciliation" on their status, and people with non-Greenlandic ethnic backgrounds started to feel even more doubtful as to whether they were included in the national "we."

In October 2014, a new election was called, for reasons unrelated to the Reconciliation Commission. Aleqa Hammond was replaced as party chairman, and it will undoubtedly prove a problem for the future of the commission that its establishment was so closely tied to Hammond's political agenda. Supposedly, the issue of full independence will also be toned down, not least due to the poor outlook for economic independence in the near future, which is a condition for political independence under the Self-Government Act.

At the same time, however, the feelings that led to the huge public support for Hammond are not likely to go away. The asymmetrical power relations between Denmark and Greenland, which still give Greenlanders a sense of being in a position of inferiority in their own country, as expressed, for example, in the debate on the continued use of Danish by the Greenlandic public, remains unresolved. Similarly, the discussion about so-called "colonial traumas" also keeps resurfacing.

In autumn 2012, Kuupik Kleist co-wrote an op-ed with a well-known and widely respected participant in the Danish–Greenlandic debate, Professor of Geology Minik Rosing, who is of Greenlandic descent (Rosing and Kleist 2012). The piece was printed in one of the national Danish newspapers and called for Danes and Greenlanders alike to preserve the narrative of their relations as a success story and to continue to pull together in exploring opportunities for Greenland's development. A key point in their article is that the narrative about the creation of modern Greenland cannot be fully separated from modern Greenland itself. If the country's formation was a failed process, it would make it difficult to see the modern country as a success. In a situation where the country needs to attract investors and to make the most of the momentum that the Arctic is currently enjoying, it is essential to present the strongest possible image. Undoubtedly, Rosing and Kleist's piece was also at least partly motivated by the fact that the Greenlandic government was getting slightly cold feet concerning its involvement with big international investors, including the Chinese. A sentence that is often heard in Greenland is that at least the Danes are familiar: "better the devil you know than the devil you don't." Thus, the commonwealth continues to provide Greenland with a safety net, both economically and mentally.

However, the two op-ed authors also addressed "the ghosts that for far too long have been fed by suppression of the truth and unsubstantiated rumours" and called for Denmark and Greenland to initiate "a clear-out of our shared lumber room" (Rosing and Kleist 2012). Lumber rooms are where we keep all the things that are not currently in use, but which we cannot bring ourselves to get rid of. Out of sight, these things continue to occupy space in our life. In combination with the term "ghosts" this conjures up an image of matters that we cannot bear to look at; matters that we may have blocked from our minds, or which may have taken on a life of their own in our memories, and which we might benefit from by exposing them to the clear light of day. These are all issues that would make topics for the type of reconciliation commission that has been established in Greenland.

SHAME, RESPONSIBILITY, AND COMING TO TERMS WITH THE PAST

What is interesting in the Greenlandic reconciliation process, as it has unfolded so far, is that it has mainly been represented as an internal process. Normally, one would expect a reconciliation to require at least two participants: the formerly opposed parties in the conflict that necessitated the reconciliation. In this case, however, one of these parties does not wish to be involved. At the yearly meeting between the three leaders of Denmark, Greenland, and the Faroe Islands in August 2013, Danish Prime Minister Helle Thorning-Schmidt underscored that this process does not reflect a Danish need, but that the Danish government fully respects the importance of this discussion for the Greenlandic people.

That is an interesting statement. If the Danish government accepts reconciliation as a legitimate Greenlandic need, it should follow that there is also a Danish need, since any relationship of course involves more than one party. The Greenlandic politicians called the Danish position regrettable.

However, the language in the Greenlandic coalition agreement does suggest that the Greenlandic initiators also view the commission primarily as a Greenlandic initiative. The reconciliation process is needed to enable the population of Greenland to come to terms with the colonial past and "put it behind them" (*"Nunatta nunasiaataasimanera qaangerniarlugu"*), as the Greenlandic version of the agreement reads. In the Danish translation, the phrase is to "dissociate from" (*"lægge afstand til"*) the colonial era. Thus, while the Greenlandic phrasing suggests acknowledgment and perhaps even a sense of catharsis, the Danish phrasing has associations with moral indignation and judgment. That both perspectives are probably in play is evident from Hammond's New Year's address and many other statements. Throughout, however, the emphasis is on the internal process. This highlights the fundamental difference between the Greenlandic and the South African commission, and underscores that the two processes do not bear comparison. The South African process was an example of so-called "transitional justice": legal processes following conflicts or oppression by the state. The idea is that this legal process needs to acknowledge the victims' rights, promote civic trust, and strengthen a democratic society based on the rule of law (International Center for Transitional Justice 2014). This is clearly about *reconciling* two (or more) parties. The Greenlandic commission, instead, is more about *reconciling with* the past, a concept that is known in German as *Vergangenheitsbewältigung*, which involves addressing and coming to terms with the past. The key point here is that legal processes need to be supplemented with an active policy of remembering, when the past is brought into light and processed. Only by initiating this sort of active remembering policy can one prevent the past from living on as unresolved traumas and unspoken feelings (Adorno 1963; Assmann and Frevert 1999; Herf 1997).[2]

If the goal is to clear out the shared lumber room of myths and prejudices and to examine the specific nature of the colonial era, it is of course regrettable if Denmark does not wish to be involved. On the other hand, it is clearly advantageous for Greenland to sit at the head of the table in this process. In connection with the Australian reconciliation process, strong criticism has been levied against a process that seemed to be more about nation-building for the majority population, which was able to affirm its own righteousness by issuing an apology, than it was about any real reconciliation with the minority population (Ahmed 2004, 101ff.).

In an excellent article, which introduces the use of Ahmed's affect theory to Danish–Greenlandic relations, Katrine Kladakis argues that Danish representations are consistently orchestrated to allow the Danes to rise above the issue and appear even more righteous by accepting the shame;

meanwhile, the shame continues to cling to the Greenlandic subjects and to Greenland's political culture (Kladakis 2012). Greenlanders live with a Danish stereotype that confines them to a dual image as either proud hunters living in harmony with nature or modern Greenlanders who failed in the transition from primitive people to modernity and therefore sink into alcohol addiction, suicide, violence, and despair. This has also become part of the Greenlandic self-concept in the widespread acceptance of a narrative about a primordial Inuit community that serves as a corrective to the modern world (Thisted 2002a). This narrative is based on the notion of the "authentic Greenlanders" whom modern Greenlanders compare themselves to. The truly authentic state is the one that existed before colonization; therefore, the more colonized, the less "authentic." When this notion is taken to its logical conclusion, dropping out of modern society becomes proof of a person's "authenticity"—hence the constant talk of a particular sense of *pride* in connection with "maladjusted" Greenlanders.

Shame is, however, the constant companion of pride, since Greenlanders live in the modern world, where the failure to cope is a source of shame. Because they are seen and also see themselves as an ethnic community, the shame is perceived collectively (Thisted 2002b). Shame is "contagious," in the sense that the shame that clings to a Greenlander who is sleeping rough rubs off on so-called "well-adjusted" Greenlanders (Rasmussen 2007; Toksvig 2010). That is why they often speak of shame, both the shame they feel and the shame they claim not to feel, yet which they know they are expected to feel and therefore feel compelled to disavow: "I have never been ashamed to be Greenlandic," says the Greenlander—thus once more cementing the link with shame. It is therefore Kladakis's point that the Greenlandic subject is already so strongly associated with shame that accepting this shame and mobilizing it as a positive transformative force simply does not seem possible—at least not within the Danish context that constitutes Kladakis's material. Of course, matters will be different when the Greenlanders define the process.

It is also debatable whether it is actually shame that the Danes accept in the representations Kladakis has analyzed, or whether they are in fact accepting a responsibility—in accordance with the colonial narrative where they were responsible for Greenland's development. Kladakis's material reflects the Danish media debate after the United Nations declared Greenland to be in violation of the Convention of the Rights of the Child in January 2009. What seemed of particular concern to some Danish politicians was the risk that Greenland's problems would reflect poorly on Denmark, qua the commonwealth. In any case, the debate clearly demonstrates how easily a Danish-led discourse about reconciliation can take on the same purpose as the reconciliation process in Australia: community-building for the majority at the cost of the minority. Someone who accepts responsibility maintains their superior position, while the one who only feels shame risks being stuck with the shame. That is also why a certain weariness has developed in Greenland towards the rhetoric of colonizer versus colonized, because it is so hard to escape a perception of the colonizer as history's active subject, while the colonized assumes the position as the passive object.

All these aspects were, no doubt, considered when the mandate for the Reconciliation Commission was drawn up. In the mandate, the decision to establish the commission was motivated by the argument that an open discussion of the past would be "beneficial for the self-awareness of the individual and of the Greenlandic people." In Greenlandic, the term *imminnut ataqqinneq* is used here; it is usually translated as "self-respect." The text implies that this self-respect is currently lacking, and the mandate thus directly considers the underlying sense of inferiority that is the precondition of Greenlandic shame. At the same time, the text seeks to avoid inappropriate victimization by suggesting that the process should be about accepting one's *own* responsibility, not about placing responsibility on others:

> The goal of the commission's efforts is to generate dialogue and insight concerning the socio-historical development in Greenland to allow us as a society to learn from the consequences of our own actions in order to create improved conditions for the future. (My translation; for the Greenlandic and Danish originals, see Saammaatta 2014)

Here, the term "own actions" allows for the sort of difficult questions that Minik Rosing and Kuupik Kleist suggest in their op-ed: What is it we think we know about the colonial era? How do we, each of us, use these narratives? And how do we move on from here?

Conclusion

The timing of the move for a Reconciliation Commission may be linked to the introduction of self-government and viewed as a desire to wipe the slate clean and move on. The need to look back is clearly stated as a desire to look forward, and thus the efforts of the Reconciliation Commission may well go hand in hand with any efforts to draw up a Greenlandic constitution. It is debatable whether the word "reconciliation," in Danish "*forsoning*," is the most apt. Both in a Danish and a Greenlandic context, the word carries connotations of abuse and blame. "*Saammaateqatigiinneq*," which is the term used in Greenlandic documents, carries profound connotations of the concept of "mercy" in the Christian religion or of the act of "pardoning" a criminal; here, however, the addition of *qatigii* underscores the mutual aspect of the process, of two parties reconciling with one another. Perhaps the word is too far-reaching, considering the actual nature of Danish–Greenlandic relations. On the other hand, when terms like "ghosts" and "lumber rooms" appear as relevant metaphors, it seems reasonable to assume that there is something at stake that warrants reconciliation.

There is an urgent need now for a new narrative to replace the two competing ones of Denmark as either the protective mother nation that took on the responsibility of developing Greenland or the exploitative imperialist nation that only had its own interests at heart. The first discourse has its roots in the nineteenth century, while the latter springs mainly from the anti-imperialism of the 1970s (Thisted 2014a). As different as these discourses may seem, they share the same underlying premise of Danes and Greenlanders belonging in separate categories and the premise of viewing the Danes as the agents of history, while the Greenlanders are cast as history's passive objects. We need a new narrative that has room for gray zones and ambiguity and, most importantly, for Greenlandic actors in the historical process.

Therefore it is crucial for the Reconciliation Commission to clarify its own mandate to ensure that it is not framed by colonial-era ideas and perspectives. In this context, it is worth carefully considering the distinction between "us" and "them." The clear-cut distinction between population groups is itself a relic of the colonial era. It was the Danish administration that divided people in Greenland into different categories, zealously controlling their mutual relations (Seiding 2013). Greenlandic sources reveal that in later years, too, it was the Danes who were most keenly interested in maintaining these distinctions. This was especially clear where the distinction was most at risk of being blurred: with respect to the educated Greenlanders (Thisted 2005). Therefore it makes much more sense to make population categories an object of the investigation rather than its premise. The commission is in no way confined by the ethnonationalist rhetoric that to some extent accompanied its establishment. On the contrary, the make-up of the commission demonstrates a desire to include the whole population, including Greenlanders with Danish as their first language as well as Danes who live and are integrated in Greenland (see the commission's website: Saammaatta 2014).

In any case, it will be one of the commission's key tasks to find a way that avoids the asymmetrical power relations, where the Danes are handed the responsibility, while the Greenlanders are left with feelings of inferiority and shame. Therefore, Greenland's determination to assume the initiative and sit at the head of the table in this process marks an important move. Denmark's decision to avoid the process altogether, on the other hand, is untenable, considering the scope of the current relations and the daily interactions between the parties. Because of the limited popular support for the commission so far, it might be expected to suffer the same fate as previous initiatives and come to nothing. However, one may also choose to hope that it will actually come up with something new and—to all parties—challenging!

NOTES

1. Greenland has a very small population, around 56,500 people, dispersed over a large number of settlements in a huge area. With home rule, *kalaallisut*, the Inuit language of the (West) Greenlanders, became the country's official language. Of the population 85 % live in an urban setting, with around 16,000 (more than a quarter of the population) living in Nuuk, the capital. The people of Thule in the extreme north-west and those of East Greenland officially accept the term *kalaallit* as the common denominator for Greenlanders, but they do not identify with the term. The residents of Thule call themselves *inughuit*, the East Greenlanders *iivit*.
2. I thank Thomas Brudholm for discussions on this issue and for very inspiring feedback and contributions in the process of writing this chapter.

WORK CITED

Adorno, Theodor W. 1963. Was bedeutet: Aufarbeitung der Vergangenheit. In *Eingriffe. Neun kritische Modelle*. Frankfurt a.M.: Suhrkamp.

Ahmed, Sara. 2004. *The cultural politics of emotion*. Edinburgh: Edinburgh University Press.

Ali, Fatuma, and Anne Lindhardt. 2006. Flugten fra Grønland. http://www.fatu-maali.dk/?Artikler. Accessed 18 Feb 2016.

Assmann, Aleida, and Ute Frevert. 1999. *Geschichtsvergessenheit—Geschichtsversessenheit. Vom Umgang mit deutschen Vergangenheiten nach 1945*. Stuttgart: Deutsche Verlags-Anstalt.

Gad, Ulrik Pram. 2009. Post-colonial identity in Greenland? When the empire dichotomizes back—Bring politics back. *Journal of Language & Politics* 8(1): 136–158.

Hammond, Aleqa. 2014. New Year speech 2014. Official translation. http://naalakkersuisut.gl/~/media/Nanoq/Files/Attached%20Files/Taler/ENG/Nytårstale%202014%20ENG.pdf. Accessed 18 Feb 2016.

Herf, Jeffrey. 1997. *Divided memory: The Nazi past in the two Germanys.* Cambridge: Harvard University Press.

International Center for Transitional Justice. 2014. What is transitional justice? http://ictj.org/about/transitional-justice. Accessed 14 Sept.

Jensen, Lars. 2012a. Nordic exceptionalism and the Nordic 'others'. In *Whiteness and postcolonialism in the Nordic region: Exceptionalism, migrant others and national identities*, ed. Lars Jensen, and Kristín Loftsdóttir, 1–11. London: Ashgate.

———. 2012b. *Danmark: Rigsfællesskab, tropekolonier og den postkoloniale arv.* Copenhagen: Hans Reitzels Forlag.

Johansen, Lars Emil. 2008. Det grønlandske folk, Det grønlandske sprog, Grønlands adgang til selvstændighed. Speech held in Nuuk, 18 June 2008.

Kladakis, Katrine. 2012. Grønlandsk Skam—Dansk Skam. Skammens Strategier i Danske Fremstillinger af Grønland. In *I Affekt. Skam, frygt og jubel som analysestrategi*, ed. Maja Bissenbakker Frederiksen and Michael Nebeling Petersen, 31–43. *Varia* no. 9. Copenhagen: University of Copenhagen, INSS, Center for Kønsforskning.

Kleist, Kuupik. 2011. Grundlovsdebatten i Grønland handler ikke særlig meget om Danmark. *Politiken*, 9 Oct 2011.

Langgård, Karen. 2011. Greenlandic literature from colonial times to self-government. In *From oral tradition to rap: Literatures of the polar North*, ed. Karen Langgård, and Kirsten Thisted, 119–188. Nuuk: Forlaget Atuagkat/Ilisimatusarfik.

Lidegaard, Mads. 1998. Grønland er ikke Sydafrika. *Information*, 4 Mar 1998.

Olwig, Karen Fog. 2003. Narrating deglobalization. Danish perceptions of a lost empire. *Global Networks* 3(3): 207–222.

Otte, Andreas Roed. 2013. Polar bears, Eskimos and Indie Music: Using Greenland and the Arctic as a co-brand for popular music. In *Modernization and heritage: How to combine the two in Inuit societies*, ed. Karen Langgård, and Kennet Pedersen, 131–150. Nuuk: Forlaget Atuagkat/Ilisimatusarfik.

———. 2014. *Popular music from Greenland. Globalization, nationalism and performance of place.* Ph.D. thesis, University of Copenhagen, Faculty of Humanities.

Petersen, Aqqaluk. 1998. Skråsikre påstande fra Fatuma Ali. *Information*, 12 Mar 1998.

Pratt, Mary Louise. 1992. *Imperial eyes: Studies in travel writing and transculturation.* London: Routledge.

Rasmussen, Inge. 2007. Interview with the Greenlandic singer Kimmernaq. *Arnanut 15*, 19 Mar 2007.

Rosing, Minik and Kuupik Kleist. 2012. Abonnement på fremtiden. *Politiken*, 20 Nov 2012.

Saammaata. 2014. http://saammaatta.gl/da/Om-os/Opgaven/Mission/Kommissorium

Seiding, Inge. 2013. *"Married to the daughters of the country": Intermarriage and intimacy in Northwest Greenland ca.1750 to 1850*. Ph.D. dissertation. Nuuk: Ilisimatusarfik.

Thisted, Kirsten. 2002a. The power to represent. Intertextuality and discourse in *Miss Smilla's sense of snow*. In *Narrating the Arctic. A cultural history of the Nordic scientific practices*, ed. Michael Bravo, and Sverker Sörlin, 311–342. Canton: Science History Publications.

———. 2002b. Som spæk og vand? Om forholdet Danmark/Grønland, set fra den grønlandske litteraturs synsvinkel. In http://tors.ku.dk/ansatte/?pure=da%2Fpublications%2Fsom-spaek-og-vand(7b32cb20-74c4-11db-bee9-02004c4f4f50).html. *Litteraturens gränsland: indvandrar- och minoritetslitteratur i nordiskt perspektiv*, ed. Satu Gröndahl, 201–223. Uppsala: Centrum för Multietnisk Forskning.

———. 2005. Postkolonialisme i nordisk perspektiv: relationen Danmark-Grønland. In *Kultur på kryds og tværs*, eds. Henning Bech and Anne Scott Sørensen, 16–43. Aarhus: Klim.

———. 2009. 'Where once Dannebrog waved for more than 200 years': Banal nationalism narrative templates and post-colonial Melancholia. *Review of Development & Change* 15(1–2): 147–172.

———. 2011. Greenlandic oral traditions: Collection, reframing and reinvention. In *From oral tradition to rap: Literatures of the polar North*, ed. Karen Langgård, and Kirsten Thisted, 63–118. Nuuk: Forlaget Atuagkat/Ilisimatusarfik.

———. 2012a. Grønland I hverdag og fest—kolonialisme, nationalisme og folkelig oplysning I mellemkrigstidens Danmark. In *Malunar Mót*, ed. Eydun Andreassen, Malan Johannesen, Anfinnur Johansen, and Turid Sigurdardóttir, 460–478. Tórshavn: Faroe University Press.

———. 2012b. Nation building—Nation branding. Julie AllStars and the act on Greenland self-government. In *News from other worlds: Studies in Nordic folklore, mythology, and culture*, ed. Merrill Kaplan, and Timothy R. Tangherlini, 376–404. Berkeley: North Pinehurst Press.

———. 2013. Discourses of indigeneity. Branding Greenland in the age of self-government and climate change. In *Science, geopolitics and culture in the polar region—Norden beyond borders*, ed. Sverker Sörlin, 227–258. Farnharn: Ashgate.

———. 2014a. Imperial ghosts in the North Atlantic. Old and new narratives about the colonial relations between Greenland and Denmark. In *(Post-) colonialism across Europe: Transcultural history and national memory*, ed. Dirk Göttsche, and Axel Dunker, 107–134. Bielefeld: Aisthesis Verlag.

———. 2014b. Cosmopolitan Inuit. New perspectives on Greenlandic literature and film. In *Globalization in literature*, ed. Per Thomas Andersen, 133–168. Bergen: Fagbokforlaget Vigmostad og Bjørke.

Toksvig, Marie Louise. 2010. Nukâka Coster-Waldau: Glem alt om udseendet. *Ekstra Bladet*, 5 Sept 2010. http://ekstrabladet.dk/flash/dkkendte/article4255700.ece. Accessed 10 Feb 2016.

CHAPTER 15

Arctic Futures: Agency and Assessing Assessments

Nina Wormbs and Sverker Sörlin

In this chapter we will examine the expanding genre of scientific assessments of the Arctic. We will focus on two examples in particular—an assessment of pollution published in 1997 entitled the *Arctic Monitoring and Assessment Programme* (AMAP) and the 2013 *Arctic Resilience Interim Report*. These two significant assessments are sufficiently far apart in time to enable us to detect change and allow for comparisons between them. We are particularly concerned with the role of science in the production of the futures that are manifested in these assessments. Given its sparse population and its key role in global environmental change, science has acquired the status of arbiter and advisor in the Arctic making it a highly desirable partner for various actors. We argue that real world interests and potential conflicts are being delegated to the scientific community, which is, willingly or unwillingly, serving as a putatively neutral and non-political quasi-authority on Arctic futures. The assessment reports therefore are not only state of the art scientific summaries, but they also give direction to the question of where the Arctic should be heading. We

N. Wormbs (✉) • S. Sörlin
Division of History of Science, Technology and Environment, KTH Royal Institute of Technology, SE 100 44 Stockholm, Sweden

© The Author(s) 2017
L.-A. Körber et al. (eds.), *Arctic Environmental Modernities*,
DOI 10.1007/978-3-319-39116-8_15

therefore analyze what happens when this question is turned over to the natural sciences, gaining political agency under the cover of "neutral" science. Assessments are almost invariably, even if only implicitly, about the future.

Assessments of the Arctic inevitably form part of the discourse in which an Arctic future is formulated, and which is central to several policy areas. Futures studies and the science of prediction have been much criticized for making far-reaching, quickly outdated, and often unwarranted assumptions about the future; hence a demand for scientific work that appears neutral with regard to prediction. There is therefore a tension between the intended purpose of the inquiry and the resulting "neutral" assessment, which anchors predicted futures in analysis in order to deduce policy recommendations from them. We propose that assessments of the Arctic constitute a specific genre that we call "future-talk," which has increasingly more discursive power in the realm of Arctic policy. For our purpose, we will not look at how assessments have actually been used in policy formulation, which was done for the *Arctic Climate Impact Assessment report* (ACIA 2004; Nilsson 2007) and requires a more comprehensive study. Rather we argue that the role science plays in these assessments is central to acquiring and maintaining discursive power. Ample previous research has demonstrated how science legitimized Arctic and Antarctic politics and policy in earlier historical periods (Sörlin 2013, 2014; Doel et al. 2014a; Dodds and Powell 2014), and how heroic deeds in remote areas, under the guise of scientific ambitions, have been a part of building nations and enhancing their identities (Herzig 2005; Hettne et al. 2006).

The large-scale "neutral" monitoring that lies behind present-day Arctic assessments is not as easily connected to policy as those hailed excursions to the frontiers of the unknown. Rather it must be understood in a context where geopolitical and environmental issues play different roles, and where resource exploitation is constantly of interest (Wormbs 2015). Science still serves national interests, but in more intricate patterns than before, mediated through issues such as resource management, sustainability, climate change, and environment, and with companies and environmental actors more immediately benefiting from the results.

ARCTIC ASSESSMENTS AS A FOCUS OF INQUIRY

The Arctic has been the focus of several scientific assessments over recent decades. These assessments present a negotiated version of state of the art science, where scientific results are filtered through a consensus process among various scientific constituencies (disciplines, institutions, commissions, organizations, etc.). They are thus not neutral vehicles of knowledge production and collection, but instead reflect preconceptions about the historical and environmental trajectory of the region and its present properties, and certain ideas regarding its future development. The concept "assessment" is in itself worth considering. An assessment is, although authored by scientists or other experts, not the same thing as a scientific paper or a scholarly book. It is commissioned work on a given topic performed during a limited time by a selected team of scientists and their aides. Literally, the word "assessment" presupposes an evaluation of the quality or performance of something. However, this definition has drifted over the course of the last two decades and an assessment is now both more and less than that. Less, insofar as there is not always an evaluation involved nor is there always a clear-cut something that can either perform or have a "quality." More, insofar as the assessment might scrutinize a range of phenomena including the states of knowledge, of policy, and of conditions and their changes.

Assessments are published as reports in different formats and contexts. In addition to scientific reports, they circulate as executive summaries, sometimes called "summaries for policy makers." The communication of assessments has undergone a kind of cultural convergence (Jenkins 2006), drawing broader social and political attention to and mobilizing around an issue. The teams producing them have become increasingly multinational, enhancing trust in the assessment undertaken. Although often evaluating some kind of performance or various processes in the present or (recent) past, they are also almost invariably preoccupied with the future, a feature that seems particularly pronounced in Arctic assessments. The concept of assessment has become part of a growing future discourse in ways that seem to be characteristic of our contemporary societies. Excellent examples of this are the Intergovernmental Panel on Climate Change (IPCC) assessments, though the trend goes back at least to the Royal Commission on Pollution reports in the UK in 1970 (Owens 2011). It is hard to find any major policy issue related to the environment or natural resources where assessments are not used as policy tools.

Dating back to the 1930s, assessments have been used in the spheres of education, psychology, and medicine, with a marked increase in their use during the 1970s and 1980s (Learned and Wood 1938; US Government 1992; Ewell 1997). An early use of assessment as a policy tool in the spheres of science and technology was by the Office of Technology Assessment, established by the United States Congress in 1972, where scientific analysis was weighed alongside stakeholder opinion and realistic policy options in a fashion that became characteristic of the assessment genre (Blair 2011). The concept of the assessment thus has an air of compromise between the scientific and the pragmatic, which is true of Arctic assessments as well. Emerging during the era of neoliberal policy evaluation schemes, the concept also has a politics. In the UK, the first Research Assessment Exercise took place in 1986 as a result of the Thatcher government's ambitious program to scale down "unnecessary" public funding to universities. It was claimed that funding should go to those that could best use it, in order to save taxpayers' money and improve the system over time by squeezing out less "productive" sectors. This process required audits, carried out across many sectors of society. A decade later this practice was common, foregrounding the idea that an "audit society" had arrived (Power 1997) or that we lived in an "audit culture" (Strathern 2000), not only in the UK, but in the OECD countries as well, where these new public management practices were implemented.

ARCTIC POLLUTION: ASSESSMENT IN THE POST-COLD WAR CONTEXT

It is against this historical backdrop that the increasing number of Arctic assessments may be seen. They can be understood as spaces of negotiation where complex problems are placed to be investigated and possible actions considered. It is therefore not surprising that assessments became a tool of the new and competitive governance structure of the Arctic that emerged at the end of the Cold War. The use of assessment reports in the Arctic was rare during the Cold War itself, when a strong nationalist and bipolar security regime held the region in a strict position. Such reports were also uncommon during the 1990s, a period characterized by peaceful and open "region building" (Keskitalo 2004). The founding of AMAP in 1991 paved the way for an increase in their use, which did not happen until the early 2000s, when the Arctic entered into a new period of increased tension and heightened controversies over actual, and more importantly, future directions of resource use, climate change impacts, policies for Arctic communities, and stakeholder status issues (Dodds and Powell 2014; Avango et al. 2013), which were behind the general increase in interest in the region and its resources.

In the so-called Rovaniemi Declaration in 1991, the Arctic Eight— Canada, Denmark/Greenland, Finland, Iceland, Norway, Russia, Sweden, and the United States—adopted the Arctic Environmental Protection Strategy (AEPS), a precursor to AMAP. The objectives of the AEPS were several and deserve to be quoted in full:

- To protect the Arctic ecosystems, including humans;
- To provide for the protection, enhancement, and restoration of environmental quality and sustainable utilization of natural resources, including their use by local populations and indigenous peoples in the Arctic;
- To recognize and, to the extent possible, seek to accommodate the traditional and cultural needs, values, and practices of indigenous peoples as determined by themselves, related to the protection of the Arctic environment;
- To review regularly the state of the Arctic environment;
- To identify, reduce, and, as a final goal, eliminate pollution. (AMAP 1997: 1)

There are five verbs in the AEPS that describe its objectives: to *protect*, *provide*, *recognize*, *review*, and *identify*. They appear to have an order where the strongest and most important is first and the least demanding is last. The final goal in the list is to eliminate pollution, but it is written in such a way that its feasibility is put into question. More important, perhaps, is the clear distinction between protecting—a word with strong agency—the ecosystems and merely recognizing or seeking to accommodate the traditions and cultural needs of humans.

In order to implement these objectives, five working groups were created: AMAP, Conservation of Arctic Flora and Fauna, Emergency Prevention, Preparedness, and Response, Protection of the Arctic Marine Environment, and Sustainable Development and Utilization. While they are equal on paper, before the release of the first extensive report in 1996 it was said that "AMAP is often viewed as the core working group from which others are supposed to base their reports and recommendations" (Russell 1996). The specific task for AMAP, outlined in the first report from 1997 entitled *Arctic Pollution Issues: A State of the Arctic Environment Report*, was to "monitor the levels and assess the effects of anthropogenic pollutants in all compartments of the Arctic environment" (AMAP 1997: vii).

Assessing, monitoring, and recommending actions have been part of the remit of AMAP from the beginning. When the Arctic Council was formed in 1996, AMAP became a working group under the Council and subsequently had an intergovernmental institution towards which to direct recommendations. AMAP bears a striking resemblance to the much more renowned IPCC (Beck 2011), with both working groups using already published scientific data in their work.

The 1997 AMAP report can be said to have two parts. The first part consisted of background information and a description of the Arctic, of the transportation of contaminants, of the ecology of the region, and of the peoples living there. In the second part, specific areas were described and the consequences of the situation were discussed, what we might call the actual assessment. This included several areas of contamination: persistent organic pollutants, heavy metals, radioactivity and acidification and haze, oil exploitation and its factual as well as possible consequences, global issues such as climate change and its effects, and how human health was impacted on by pollution. The report also contained an executive summary with recommendations. Even though the report can be divided into two categories, the difference between the descriptive chapters and the assessing ones is very small, if detectable at all. The entirety of the text is descriptive without analysis. In accordance with standard scientific language usage, value-laden concepts and reasoning are omitted, and issues are stated as a matter of fact. Surprisingly, this also goes for the executive summary in which recommendations for political action were given. Only a few recommendations involved informing indigenous peoples, while most aimed at securing scientific and other knowledge as the basis for action. The report stated that international strategies needed more information, better-developed models, and long-term monitoring over a larger area to ensure greater coverage and to meet local needs. With regard to policy suggestions, the overarching message was to develop and adhere to international laws and protocols (AMAP 1997: xi–xii).

A number of other reports followed the inaugural AMAP report in the form of either updates or new topics. During this period, AMAP came under the auspices of the Arctic Council, which specifically requested

AMAP to "recommend actions required to reduce risks to Arctic ecosystems" (AMAP 2002: iv). A shift in language can be identified, where words such as "evidence" and formulations such as "been established" signal a broader knowledge base. The suggestions for action continued to focus on the need for new knowledge, even though there were discrepancies in terms of what knowledge was deemed important in order to raise certain questions and continue monitoring. This is illustrated by an AMAP report in 2012 on the cryosphere, otherwise known as SWIPA (Snow, Water, Ice, Permafrost in the Arctic). Here, very specific questions regarding ice melt and the effect of an increasing amount of freshwater in ecosystems were put next to the broader, and indeed daring, question of how the changes will "affect Arctic societies and economies" (AMAP 2012: x). A general assumption in science-based assessments is that the changes in ecosystems and other natural systems that science can identify are readily studied as the chief root causes ("drivers") of change in societies, although the relative explanatory power of these is not particularly well addressed. This is a peculiar feature of the assessments we have studied.

Arctic Resilience: Assessment in the Anthropocene?

An important milestone in the production of Arctic assessments was the encompassing ACIA in 2004. Still based on a natural science understanding of knowledge production, it managed to integrate indigenous people in the process, framing certain issues more broadly and including elements of social science (Nilsson 2007).

The Arctic Resilience Interim Report (ARR 2013; a final report is expected in late 2016) is our second main example. The ARR should be analyzed in view of the extended scope that the assessments acquired over time. Even though it was an Arctic Council project, the ARR was not produced within the classic framework of AMAP, but rather departs explicitly from a theoretical resilience framework and from the related theory of social-ecological systems. Instead ARR emphasizes methodological and framing assumptions, conceptual deliberations, and discussions of limits and potentials in the chosen perspectives. No resilience assessment has ever been conducted on such a large scale before. The report begins with a long, critical, and informative discussion about some of the key assumptions in resilience theory, for example the idea that systems reach critical thresholds or tipping points at which they change states or regimes. While empirically found to be the case in local ecological systems, the empirical evidence is far less convincing when it comes to societies or even subsets of societies, let alone when combinations of ecosystems and social systems are considered. If it is assumed that ecological and social systems are fundamentally interconnected, based on the literature review in the third chapter of the ARR, it still seems undetermined whether this interconnectedness actually bears out at a pan-Arctic scale.

The operational demands on the report have made some limitations necessary. The resulting omissions speak to the kind of results that seem conceivably possible to achieve, even in a very ambitious undertaking such as the ARR. To a great degree, this has to do with the kind of methodological choices that are made, which in turn are predicated both on the resilience perspective and on the context of the report and its remit from the Arctic Council. The report displays a clear understanding of both. It states explicitly that resilience ultimately rests on choice—of what system(s) should be resilient and how this choice in turn rests on values and therefore on governance. The report also uses the concept of power in this respect: "from the governance perspective, resilience reflects the desires of those with the power to make and implement decisions" (Robards et al. 2011: 22). Even with this high level of awareness, the conditions framing the report create an ambivalent relationship to certain kinds of choices. A common denominator of these choices is how effectively they avoid engaging with agency. The language of "drivers," inherited from the earlier, exclusively science-based, assessments, seems in this regard to be a constraint, although the range of drivers is here expanded to include societal ones.

In the section on thresholds, the report presents two illustrations of drivers, one social, the other biophysical, which are distributed in relation to their timescales (Figs. 15.1 and 15.2). Figure 15.2 identifies a range of perfectly obvious and undisputable factors that are effecting Arctic change, but does not include factors related to social agency. There is no mention of politics, social movements, or ideas of any kind, such as political ideologies. If Arctic change—from a resilience perspective—is an integrated change across the social and the biophysical, and across "multiple temporal and spatial scales" as is repeatedly underscored in the report, why are such fundamental categories in an analysis of change excluded from serious analysis? They show up as statements, but are not engaged with. (This is also true for a few figures.) One answer would be that it is too complicated, that the ARR is already dealing with something very complex, which the caption alludes to in its evasive language on the mechanism of

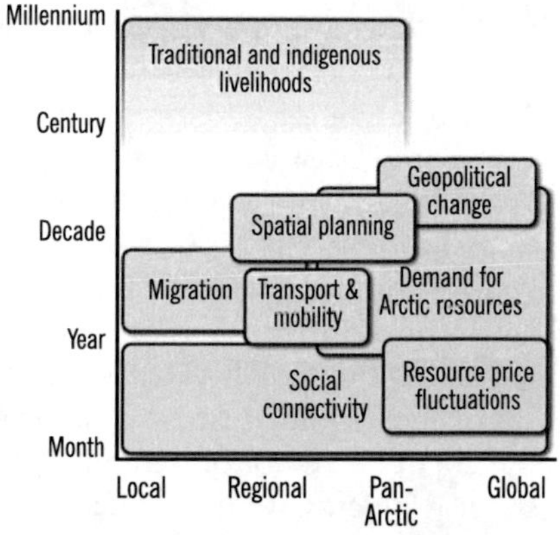

Fig. 15.1 Figure from the *Arctic Resilience Interim Report* (2013). Graphics credits: Hugo Ahlenius, Nordpil

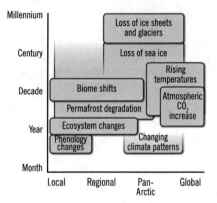

Fig. 15.2 Figure from the *Arctic Resilience Interim Report* (2013). Graphics credits: Hugo Ahlenius, Nordpil

change. On the other hand, the approach does require the identification of drivers, because, if these are not identified, then the change that is absolutely crucial for the analysis would be limited to the past and the present and could say nothing about the future. Drivers are necessary for the kind of assumed directionality that makes this publication an assessment rather than merely a study of the Arctic "as is."

The critical moment of the assessment is when it turns from the descriptive to the prospective. How does it do that? If agency is an important dimension, one would expect to see reflexive work on the identified social drivers. Questions about if and how change might happen in geopolitics, ownership, resource demand, resource prices, "social connectivities," and social planning would be raised. This discussion seems more or less futile if the underlying political and ideological values are not factored in, and if agency in society is not counted upon. The drivers identified in the ARR (2013: 44–45) are presented in detail in a chart and are described using neutral words such as geopolitical change, globalization, migration, and increased resource demand. Another category called "Observed change in the Arctic" specifies the changes, using words and phrases like militarization, urbanization, opening of trans-polar shipping routes, and financial investment. A third category explains why these changes are important. The fourth category lists some key references for all dimensions of the identified Arctic change drivers. It should be pointed out that these changes are not presented as desirable or idyllic. In a few instances, distinct problems are mentioned, such as the risk of the loss of biodiversity and of traditional knowledge, but in other instances opportunities are underscored. The language on why something is a risk or an opportunity, or a "positive feedback," isn't very developed. Change is sometimes good and sometimes bad, often depending on who is asked. But what does it all have to do with resilience?

Because the ARR does not engage with values, ideas, politics, social movements, or other dimensions of agency, nor connect the understanding of the predications of the resilience approach with a discussion of the potential effects of the drivers, the answer is largely lacking. To privilege agency would imply the necessity of taking a stance on the issue of "resilience for what?" Such questioning would turn the report into something more explicitly value-laden, or even political. One way around such a turn toward the political is to avoid siding explicitly with any kind of desired future. In the absence of such engagement with "real" values-based resilience, the de facto political role played by the ARR, drawing on its interim report, may be the affirmation and tacit confirmation of the current orthodoxy, which reproduces established power structures, regardless of whether or not they seem capable of sustaining resilience at all.

A Theory of Arctic Assessments?

It should be stressed that there are few other regions in the world where scientific assessments have the position they have in the Arctic. It is likely that no other region of the world has had as many assessments per capita. The limited number of interests can more or less be "handled" through the implicit agency allowed for by the assessments. A similar monitoring and assessing of a populated region would be unthinkable because the "future" cannot be as neatly delimited. Not only are Arctic assessments different from assessments in the rest of the world, but they also differ amongst themselves. In the two key reports analyzed here, the agency attributed is distinctly different. In the AMAP pollution report from 1997, change is inherent in the description of the monitoring, and is largely isolated to slow changes in nature and ecosystems. In the 2013 ARR the identified drivers are the cause of change. The entire structure of the ARR is more complex and attempts to allow for a broader analysis, sometimes also called the co-production of nature and society, whereas in the AMAP report, societal change is marginal to the analysis. The ARR of 2013 is only an interim report, but it nonetheless demonstrates that the analysis of societal consequences and deliberations must be at the core of the assessment. In practice, the attempt to do so does not lead very far, however, not least because the methodological and theoretical framework of resilience theory restricts the potential use of knowledge from social science and the humanities.

Assessments therefore function as a de facto affirmation of a future-oriented discourse, taking as valid points of departure actual trends and tendencies surrounding Arctic change (see Emmerson 2010). They confirm the existence of futures by using them as analytical drivers and causal factors in the reports. This is clearly visible in the ARR, which lists these drivers and uses them as the analytical framework of the study. The question the analysis tries to answer appears to be the following: These are the future trends, and, given such a future, is the Arctic resilient?

All of the reports we studied generally affirm what can be called a hegemonic Arctic discourse. Hegemony, in a Gramscian sense, is a consensual understanding among societal elites (Cox 1993; Gill 1993), a concept that has recently been applied to the discourse of Arctic geopolitics (Ahearne 2013; Hough 2013). It is clear that the "assessment industry" in the Arctic is affected by this hegemony and is, in practice, enrolled in it. All assessment reports avoid an engaged, critical reflection about the very parameters of their own undertaking. Reports offer few or any alternative ways of thinking and they deliberately avoid a theoretical basis that would allow for critical reflection. These thus actively abstain from questioning the hegemonic discourse and from suggesting anything that would resemble the Gramscian concept of counter-hegemony. In this respect, these assessments should not surprise us. There is widespread understanding in the literature on scientific advice that assessments tend to work on the premises set by those who commission them. The way the analysis is set up privileges quantitative data and scientific methods, and marginalizes and/ or flattens culture, history, society, and agency; such approaches are consistent with how the so-called "human dimension" has been treated in climate change and sustainability studies thus far (Hulme 2011; Castree et al. 2014). This trend has been sustained by stereotypical media narratives of climate change (Boykoff 2011) and their impact in the Arctic (Christensen et al. 2013). The Arctic assessments we have studied conform well to what has been called the "linear model of scientific expertise." Such linear models tend to operate with an understanding of expertise as a science, and its main mission is to "get the science right" (Beck 2011).

This is not the only position that scientists can take, however. Recent literature on science advice has argued that the linear model of expertise actually limits the use of the scientific expertise, privileging the natural sciences, which have an object of study and a spectrum of expertise from the knowledge that underlies the policy work itself. Scholars have also argued that the linear model in and of itself implies a tacit politicization of the scientists' role in advice (Elzinga 1996; Jasanoff and Wynne 1998; Pielke 2007; Sarewitz 2010). Yet, as many scholars in the field of science and technology studies have shown, scientists and scholars should engage deeper in the politics of science and advice in order to sustain its independence and provide more realistic and useful recommendations (Miller and Edwards 2001; Pielke 2007; Sarewitz 2000). The possibility of engaging more intently with the politics of science advice has not yet been actualized very actively in Arctic assessments. It could be very different if a wider range of expertise would mobilize and undertake a more reflexive mode of engagement. Thus, the scientists play a perhaps inadvertent role in tacitly endorsing a certain set of future interests in the Arctic.

Work Cited

ACIA. 2004. *Arctic climate impact assessment – Scientific report.* Cambridge: Cambridge University Press.

Ahearne, Gerard. 2013. Towards an ecological civilization: A Gramscian strategy for a new political subject. *Cosmos and History: The Journal of Natural and Social Philosophy* 9(1): 317–326.

AMAP. 1997. *Arctic pollution issues: A state of the Arctic environment report.* Oslo: AMAP.

———. 2002. *Arctic pollution 2002: Persistent organic pollutants, heavy metals, radioactivity, human health, changing pathways.* Oslo: AMAP.

————. 2012. *Arctic climate issues 2011: Changes in Arctic snow, water, ice and permafrost. SWIPA 2011 overview report.* Oslo: AMAP.

ARR. 2013. *Arctic resilience: Interim report 2013.* Stockholm: Stockholm Environment Institute and Stockholm Resilience Centre.

Avango, Dag, Annika E. Nilsson, and Peder Roberts. 2013. Assessing Arctic futures: Voices, resources, governance. *Polar Journal* 3(2): 431–446.

Beck, Silke. 2011. Moving beyond the linear model of expertise? IPCC and the test of adaptation. *Regional Environmental Change* 11: 297–306.

Blair, Peter D. 2011. Scientific advice for policy in the United States: Lessons from the National Academies and the Former Congressional Office of Technology Assessment. In *The politics of scientific advice: Institutional design for quality assurance,* ed. Justus Lentsch, and Peter Weingart, 297–333. Cambridge: Cambridge University Press.

Boykoff, Max T. 2011. *Who speaks for the climate? Making sense of media reporting on climate change.* New York: Cambridge University Press.

Castree, Noel, William M. Adams, John Barry, Daniel Brockington, Bram Büscher, Esteve Corbera, David Demeritt, Rosaleen Duffy, Ulrike Felt, Katja Neves, Peter Newell, Luigi Pellizzoni, Kate Rigby, Paul Robbins, Libby Robin, Deborah Bird Rose, Andrew Ross, David Schlosberg, Sverker Sörlin, Paige West, Mark Whitehead, and Brian Wynne. 2014. Changing the intellectual climate. *Nature Climate Change* 4(9): 763–768.

Christensen, Miyase, Annika E. Nilsson, and Nina Wormbs (ed). 2013. *Media and Arctic climate politics. Breaking the ice.* New York: Palgrave Macmillan.

Cox, Robert W. 1993. Gramsci, hegemony, and international relations: An essay in method. In *Gramsci, historical materialism, and international relations,* ed. Stephen Gill, 49–66. Cambridge: Cambridge University Press.

Dodds, Klaus, and Richard C. Powell (ed). 2014. *Polar geopolitics: Knowledges, resources and legal regimes.* Cheltenham: Edward Elgar.

Doel, Ronald E., Robert Marc Friedman, Julia Lajus, Sverker Sörlin, and Urban Wråkberg. 2014a. Strategic Arctic science: National interests in building natural knowledge – Interwar era through the Cold War. *Journal of Historical Geography* 42: 60–80.

Doel, Ronald E., Urban Wråkberg, and Suzanne Zeller. 2014b. Science, environment, and the New Arctic. *Journal of Historical Geography* 42: 2–14.

Elzinga, Aant. 1996. UNESCO and the politics of international cooperation in the realm of science. In *Les sciences coloniales 2,* ed. Patrick Petitjean, and Roland Waast, 91–132. Paris: Orstrom.

Emmerson, Charles. 2010. *The future history of the Arctic.* London: Bodley Head.

Ewell, Peter T. 1997. Accountability and assessment in a second decade: New looks or same old story? In *Assessing impact, evidence and action,* 7–22. Washington, DC: American Association of Higher Education.

Foucault, Michel. 2002. *The archaeology of knowledge and the discourse on language.* Trans. A. M. Sheridan Smith. London/New York: Routledge. Originally published as Michel Foucault. 1969. *L'archéologie du savoir.* Paris: Gallimard.

Gill, Stephen (ed). 1993. *Gramsci, historical materialism, and international relations.* Cambridge: Cambridge University Press.

Herzig, Rebecca. 2005. *Suffering for science: Reason and sacrifice in modern America.* New Brunswick: Rutgers University Press.

Hettne, Björn, Sverker Sörlin, and Uffe Østergård. 2006. *Den globala nationalismen: Nationalstatens historia och framtid,* 2nd ed. (orig. 1998). Stockholm: SNS förlag.

Hough, Peter. 2013. *International politics of the Arctic: Coming in from the cold.* Abingdon: Routledge.

Hulme, Mike. 2011. Meet the humanities. *Nature Climate Change* 1: 177–179.

Jasanoff, Sheila, and Brian Wynne. 1998. Science and decision making. In *Human choice and climate change,* ed. Steve Rayner, and Elizabeth L. Malone, 1–87. Columbus: Battelle.

Jenkins, Henry. 2006. *Convergence culture: Where old and new media collide.* New York: New York University Press.

Keskitalo, E.Carina.H. 2004. *Negotiating the Arctic: The construction of an international region.* New York: Routledge.

Learned, William S., and Ben DeKalbe Wood. 1938. *The student and his knowledge.* New York: Carnegie Foundation for the Advancement of Teaching.

Miller, Clark A., and Paul N. Edwards. 2001. *Changing the atmosphere: Expert knowledge and environmental governance.* Cambridge, MA: MIT Press.

Nilsson, Annika E. 2007. *A changing Arctic climate: Science and policy in the Arctic climate impact assessment.* Linköping: Linköping University.

Owens, Susan. 2011. Knowledge, advice and influence: The role of the UK Royal Commission on Environmental Pollution, 1970–2009. In *The politics of scientific advice: Institutional design for quality assurance,* ed. Justus Lentsch, and Peter Weingart, 73–101. Cambridge: Cambridge University Press.

Pielke, Roger A. 2007. *The honest broker: Making sense of science in policy and politics.* Cambridge: Cambridge University Press.

Power, Michael. 1997. *The audit society: Rituals of verification.* Oxford: Oxford University Press.

Robards, Martin D., Michael L. Schoon, Chanda L. Meek, and Nathan L. Engle. 2011. The importance of social drivers in the resilient provision of ecosystem services. *Global Environmental Change* 21(2): 522–529.

Russell, Bruce A. 1996. The Arctic environmental protection strategy and the New Arctic Council. *Arctic Research of the United States* 10: 2–8. http://arcticcircle.uconn.edu/NatResources/Policy/uspolicy1.html

Sarewitz, Daniel. 2000. Science and environmental policy: An excess of objectivity. In *Earth matters: The earth sciences, philosophy, and the claims of community*, ed. R. Frodeman, 79–98. Upper Saddle River: Prentice Hall.

———. 2010. Normal science and limits on knowledge. *Social Research* 77(3): 997–1010.

Sörlin, Sverker (ed). 2013. *Science, geopolitics and culture in the polar region: Norden beyond borders*. Farnham: Ashgate.

———. 2014. Circumpolar science: Scandinavian approaches to the Arctic and the North Atlantic, ca. 1930 to 1960. *Science in Context* 27(2): 275–305.

Strathern, Marilyn. 2000. *Audit culture: Anthropological studies in accountability, ethics and the academy*. London: Routledge.

U.S. Government. 1992. Department of Education, National Center for Education Statistics. *National assessment of college student learning: Issues and concerns*. Washington, DC: U.S. Government Printing Office.

Wormbs, Nina. 2015. The assessed Arctic: How monitoring can be silently normative. In *The new Arctic*, ed. Birgitta Evengard, Joan Nyman Larsen, and Øyvind Paasche, 291–301. New York: Springer.

INDEX

© The Author(s) 2017   263
L.-A. Körber et al. (eds.), *Arctic Environmental Modernities*,
DOI 10.1007/978-3-319-39116-8